American Book Company

The Standards Experts

ACT® Science Test Preparation Guide

Authors: Liz Thompson
Michelle Gunter
Emily Powel
Clive Sombe

American Book Company
PO Box 2638
Woodstock, GA 30188

Toll Free: 1 (888) 264-5877 Phone: (770) 928-2834
Fax: (770) 928-7483 Toll Free Fax: 1 (866) 827-3240
www.americanbookcompany.com

Acknowledgements

The authors would like to gratefully acknowledge the formatting and technical contributions of Becky Wright.

We also want to thank Mary Stoddard and Eric Field for their expertise in developing the graphics for this book.

A special thanks to Marsha Torrens for her editing assistance.

PO Box 2638
Woodstock, GA 30188-1383

Printed in the United States of America

04/17

American Book Company

The Standards Experts

Dear Student

Welcome to your new American Book Company Learning Tool!
This book has been created especially for you. Our writers have covered 100% of the standards and concepts as clearly and simply as possible.

We look forward to hearing of your success!

G E A R

GAUGE

Pre-tests help educators createan individualized learning plan for every student.

EDUCATE

Abundant lessons and practice exercises promote student mastery. In-book or online testing allows students to practice the skills they have learned in a test environment.

APPLY

Students approach their exams with new self-confidence and reinforced skills.

RESULTS

Our programs have been proven to prepare students for College and Career Readiness (CCR) and increase performance.

American Book Company's GEAR Learning Programs

We ensure students understand and can apply the knowledge needed to increase their performance.

Each GEAR Learning Program blends both text and digital resources, including student workbooks and ebooks for student use or classroom display!

Table of Contents

Preface

The *ACT Science Test Preparation Guide* will help students preparing to take the ACT Science Test. **The materials in this book were developed by the American Book Company.** The passages and questions were designed to simulate the style of those found on the ACT©. The ACT is developed and prepared by the American College Testing Program. The American College Testing Program had no role in the development of this text.

This book contains several sections: 1) general information about the book, 2) a complete science reasoning test, 3) an evaluation chart, 4) chapters that review the concepts and skills that improve readiness for the ACT Science Reasoning Test and 5) two complete practice science reasoning tests for additional practice. Answers to the tests and chapter reviews are in a separate manual.

We welcome comments and suggestions about the book. Please contact the authors at

American Book Company
PO Box 2638
Woodstock, GA 30188-1383

Toll Free: 1 (888) 264-5877
Phone: (770) 928-2834
Fax: (770) 928-7483
Website: www.americanbookcompany.com

About the Authors

Liz A. Thompson holds a B.S. in Chemistry and an M.S. in Analytical Chemistry, both from the Georgia Institute of Technology. Research conducted as both an undergraduate and graduate student focused on the creation and fabrication of sensors based on conducting polymers and biomolecules. Postgraduate experience includes work in radioanalytical chemistry. Her publications include several articles in respected scientific journals, as well as partial authorship of the textbook *Radioanalytical Chemistry* (2007). At every educational level, Mrs. Thompson has enjoyed teaching, tutoring and mentoring students in the study of science.

Michelle Gunter graduated from Kennesaw State University in Kennesaw, Georgia, with a B.S. in Secondary Biology Education. She is a certified teacher in the field of Biology in the state of Georgia. She has three years experience in high school science classrooms. She has nine years experience in biology and biological systems. She has won awards for her research in the field of aquatic toxicology. Mrs. Gunter enjoys teaching students of all ages the wonders of the natural world.

Emily J. Powell holds a B.S. and an M.S. in Physical Geography with an emphasis in atmospheric science, both from the University of Georgia. Her graduate research focused on the surface mass balance of the Greenland ice sheet, comparing ERS satellite and model data. She holds partial authorship of a published research paper in a respected meteorological journal. She has five years experience in meteorology product development. She has two years experience teaching and tutoring students of all ages in classroom and professional environments in both the United States and Spain.

Clive Sombe received most of his education from his home country, Zambia and from The United Kingdom. Clive holds an Advanced Diploma in Management Accounting as well as A-Levels in Mathematics, Chemistry and Physics from Cambridge University. Currently, Clive is an honor student at Kennesaw State University and is expected to graduate with a B.S in Mathematics this fall. He enjoys tutoring, writing and spends some of his spare time exploring mathematical trivia.

Diagnostic Test

SCIENCE TEST

35 Minutes—40 Questions

DIRECTIONS: There are seven passages in this test. Each passage is followed by several questions. After reading a passage, choose the best answer to each question. You may refer to the passages as often as necessary.

You are ***NOT*** permitted to use a calculator on this test.

Passage I

Omega-3 is a type of fatty acid animals need for proper neural development and lifelong health. This fatty acid must be consumed in the diet of all mammals. Until recently, one of the few sources of omega-3 was fish oils, which also contain harmful heavy metals. Researchers at the University of Missouri have genetically inserted a gene, the fat-1 gene, from a species of roundworm, into pigs. This gene makes the engineered pigs produce omega-3 fatty acids in lieu of harmful saturated fats.

Two doctors were discussing the consequences of creating a bioengineered pig.

Doctor 1

Creating a healthier pig for consumption benefits all mankind. The creation of this terrestrial animal takes pressure off depleted fisheries around the world. In addition, pigs do not contain dangerous levels of metals found in fish. Creating a healthier pig makes the pigs themselves healthier, leading to increased lifespan and decreased costs to pig farmers. Pigs readily consume any manner of food and grow to slaughtering weight before their first birthday. In addition, female sows produce almost 20 piglets a year. The production of cheap, healthy and tasty agricultural crops has been the goal of farmers for many years. The only difference between ancient domestication practices and modern engineering is the rate at which humans alter animals.

Doctor 2

Creating a pig with altered DNA is unnatural, and humans cannot possibly begin to understand the ramifications of such actions. Genetically modified foods can cause potential allergies in some humans. A food that was once considered safe may now contain proteins that could elicit fatal allergic responses in humans. Also, this food product could escape captivity and begin living in the wild. The increased lifespan of the omega-3 pigs means female sows will live longer in the wild, producing more litters. Female sows can produce around 20 piglets a year. This further compounds an already exacerbated invasive species situation. As evidenced in Texas, Wisconsin, Georgia and many other US states, non-native feral pigs are highly destructive to lands.

1. **According to Doctor 2, the creation of omega-3 pigs:**
   - A is beneficial to mankind.
   - B is harmful to mankind.
   - C is benign to mankind.
   - D is unavoidable.

2. **What two concerns most likely led scientists to develop an omega-3 pig?**
   - F Heavy metals found in fish and over-fishing of the oceans
   - G The high-fat American diet and high incidence of heart disease
   - H Helping people with food allergies and creating cheap foods
   - J Large feral pig population and the unwillingness of humans to eat fish

3. **What unforeseen benefit does the insertion of the fat-1 gene have on the domestic pig population?**
   - A Longer lives
   - B Larger litters
   - C Shorter legs
   - D Larger ears

4. **According to Doctor 1, there is a negative relationship between the level of consumption of omega-3 pork products and the development of age-related illnesses such that:**
   - F as you consume more omega-3 pork products, you develop more age-related illnesses.
   - G as you consume more omega-3 pork products, you develop fewer age-related illnesses.
   - H as you consume fewer omega-3 pork products, you develop fewer age-related illnesses.
   - J as you consume fewer amounts of omega-3 pork products, you develop more age-related illnesses.

5. **According to Doctor 1, how long have humans been altering the look and function of animals?**
   - A Since ancient times
   - B Only in the past 500 years
   - C Only in the past 50 years
   - D Only in the past 5 years

6. **How does Doctor 1's viewpoint differ from Doctor 2's viewpoint?**
   - F Doctor 1 believes that risky, genetically altered food is beneficial to mankind, while Doctor 2 asserts that genetically altered food is very beneficial and free of any risk.
   - G Doctor 1 believes that genetically engineered food products will help people, while Doctor 2 believes genetically engineered foods are more harmful than helpful.
   - H Doctor 2 believes that genetically engineered food products will help people, while Doctor 1 believes genetically engineered foods are more harmful than helpful.
   - J Both Doctor 1 and Doctor 2 are ambiguous in their argument.

7. **According to the passage, why are omega-3 fatty acids important to consume?**
   - A They are found in fish, an important economic crop.
   - B They play a critical role in neural development.
   - C They actively reduce body fat in humans.
   - D They form the foundation of bone growth in infants.

Passage II

Human cognitive physiologists define *intelligence* as: "the ability to learn or to deal with new or trying situations." For the last 100 years, the scientific world has viewed animals as "machines" that simply complete tasks based on inherited innate behaviors. It has been a long standing tenet of the scientific community that only humans possess cognitive abilities. New research studies are challenging this historical scientific theory.

Crows were presented with the circumstance pictured in Figure 1. A piece of food was placed in a basket, and the basket was placed in a long glass tube. The crow was given a material it could use to fashion into a tool and was left alone to solve the problem. If the crow could fashion the material into an appropriate tool, it could reach the food.

Several experimental trials were conducted using American Crows (*Corvus brachyrhynchos*) as the subjects.

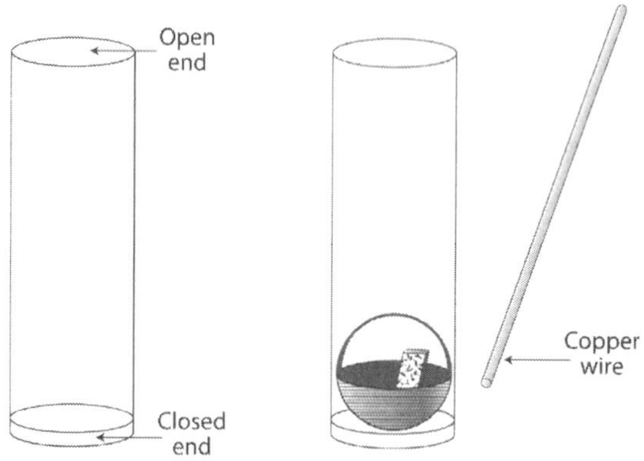

Experiment 1

An American Crow named Betty is given copper wire. The time it takes her to shape this material into a tool capable of reaching the food reward is recorded in Table 1.

Table 1		
Betty		
Trial Number	Material Tested	Completion Time
1	Copper Wire	121 seconds
2	Copper Wire	96 seconds
3	Copper Wire	65 seconds
4	Copper Wire	33 seconds

Experiment 2

In this experiment, Betty is given a variety of materials with which to make a tool.

Table 2	
Betty's Average Trial Time	
Material	Average Time
Stick	44 seconds
Vine	61 seconds
Spoon	72 seconds
Copper wire	79 seconds
Aluminum Foil	90 seconds

In the third experiment, four different crows are given copper wire and timed.

Table 3			
Other Data			
Bird Name	Material Selected	Time	Temperature of Room
Betty	Copper wire	22 seconds	77°C
Carl	Copper wire	33 seconds	78°C
Mike	Copper wire	91 seconds	77°C
Jason	Copper wire	37 seconds	76°C

8. According to Table 1, how did Betty's completion time change between Trials 3 and 4?

F She completed Trial 4 in 1/2 the time of Trial 3.

G She completed Trial 4 in twice the time of Trial 3.

H She completed Trial 3 and Trial 4 in the same time.

J She completed Trial 3 in 1/3 the time of Trial 4.

9. Betty's average time completing this experiment using a wire made out of copper would be closest to which value?

A 110 seconds

B 80 seconds

C 60 seconds

D 20 seconds

10. What useful information does Table 3 show about the experiment?

F How temperature relates to tool choice

G How speed of response relates to temperature

H How tool choice relates to speed of response

J How individual skill relates to speed of response

11. Table 2 shows the crow's response versus time, using a variety of materials. Which statement best describes their response?

A Crows work best with metallic materials.

B Crows work best with organic materials.

C Crows show a preference for man-made items.

D Crows work equally well with all materials.

12. The main purpose of this experiment was:

F to prove the animals as "machines" theory.

G to show that animals are not intelligent.

H to show that birds are intelligent.

J to demonstrate the problem-solving abilities of crows.

13. According to Table 1, which of the following graphs best illustrates the relationship between trial number and time to completion?

A

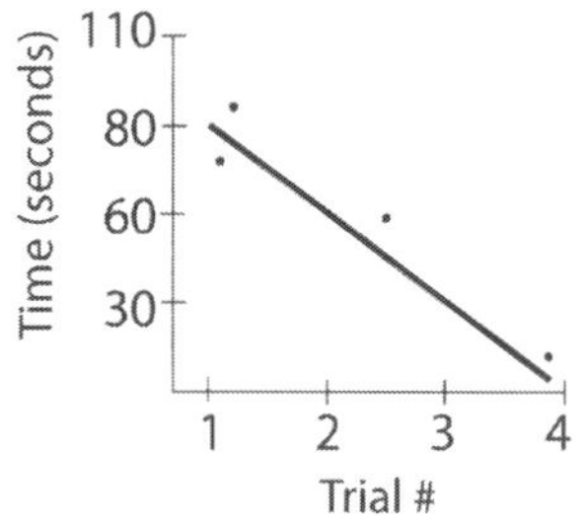

B

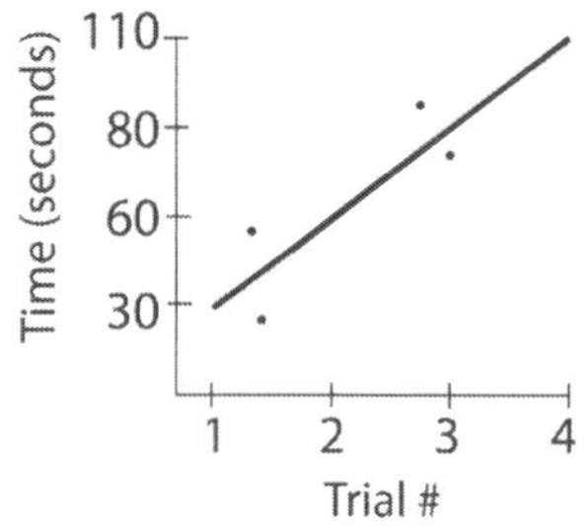

C

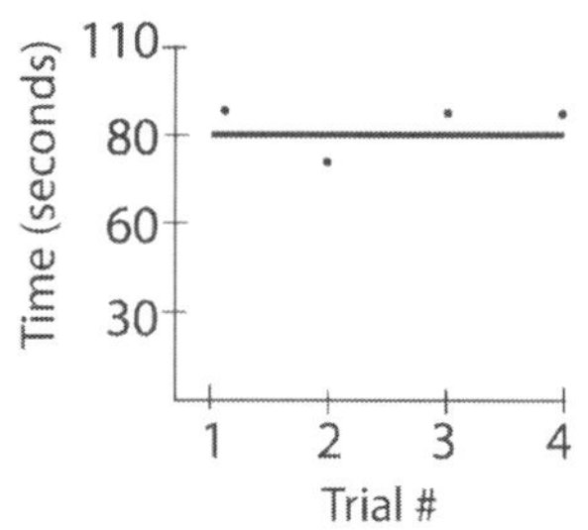

D

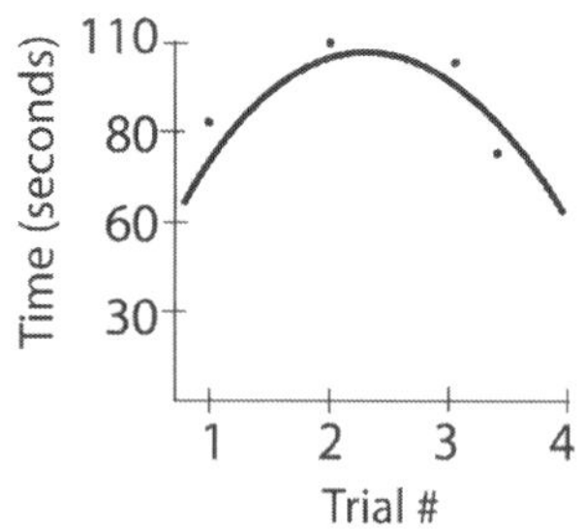

Passage III

A *mesoscale convective complex* (MCC) is a grouping of thunderstorms organized on a larger scale than individual storm cells. These significant convective systems produce severe weather and precipitation that impact much of the United States. They contribute a large amount of precipitation during the growing season, particularly in the central and eastern United States. The elimination of MCCs would greatly change the water cycle in these regions.

A study collected data on MCCs for a 15-year period between 1988 and 2009. The study included 527 events, for an average of 35 MCCs per year. Figure 1 shows the average number of MCCs per month for this 15-year period. Figure 2 shows average MCC densities for four months of the year, May (a), June (b), July (c) and August (d). The densities were calculated using the number of hours an MCC cloud shield remained over an area during the storm's lifetime. Higher numbers indicate the storm lasted a greater number of hours.

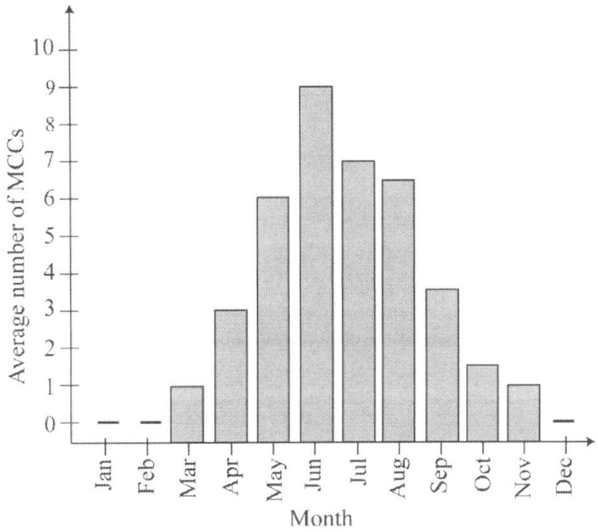

Figure 1

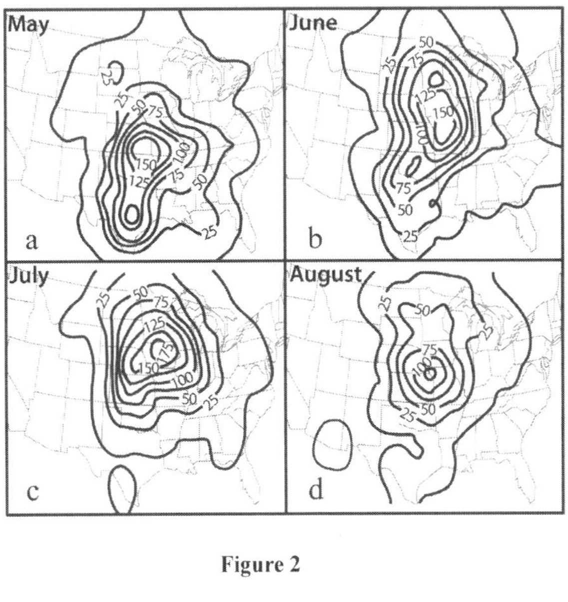

Figure 2

14. MCCs typically peak in June. According to Figure 1, what is the average number of events occurring per year during this month?

F About 5

G About 9

H About 10

J Exactly 7.5

15. Using Figure 1, during which months do MCCs contribute most to precipitation?

A April, May, June, July

B September, October, March, April

C June, July, August, September

D May, June, July, August

16. Using Figure 2, in what month is the duration of MCCs the greatest on average?

F May

G June

H July

J August

17. Using Figure 1, which of the following statements best describes the pattern of MCC events?

A They occur only in the warm season with no clear peak period.

B They occur during warm and cool months but peak in summer.

C They can occur at any time of the year but peak in April and August.

D They peak in cool months and last throughout the warm season.

18. From May to June, how does the core of MCC densities change according to Figure 2?

F The density core shifts towards the northeast into Missouri.

G The density core shifts towards the southwest into Texas.

H The density core shifts towards the northwest into Nebraska.

J The density core shifts toward the southeast into Georgia.

Passage IV

Scientists have been researching maglev trains since the 1950s. The name is derived from magnetic levitation and is a system of train transportation that uses magnetic levitation. The trains are suspended above the track, guided on the track and propelled forward solely by electromagnetic fields. The electromagnetic fields are generated by either electromagnets or permanent magnets.

An *electromagnet* is a device that becomes magnetic when an electric current flows through it. An electromagnet only generates a magnetic field as long as current flows. As soon as the electricity is turned off, an electromagnet is no longer magnetic.

A *permanent magnet* remains magnetic unless its structure is disrupted by heating or physical damage.

Scientists have developed two primary types of maglev technology.

Technology 1

The electromagnetic suspension (EMS) system uses attractive magnetic forces to lift the train up. The EMS system uses electromagnets. Electromagnets are attached to the train and oriented toward the steel rail from below. The degree of the attraction between the electromagnet and the rail is modulated by the current applied to the electromagnet. The electromagnets use feedback control to maintain a train at a constant distance from the track.

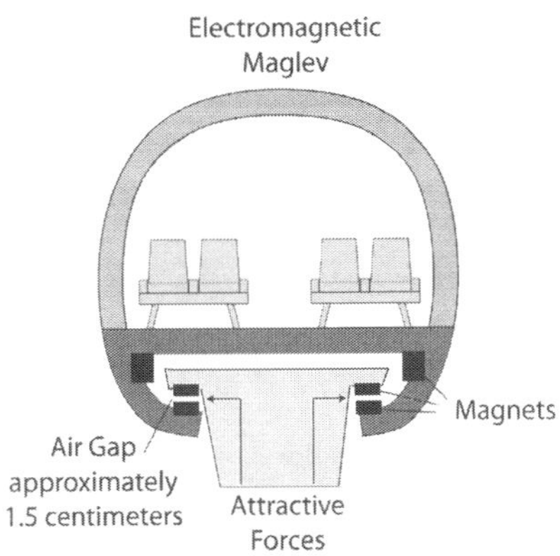

Figure 1

Technology 2

The electrodynamic suspension (EDS) uses repulsive magnetic forces to lift the train up. In this system, both the train and the rail contain a magnet. The poles of two magnets are aligned to repulse one another. The EDS system features trains that can have either electromagnets or permanent magnets. The rail contains an electromagnet.

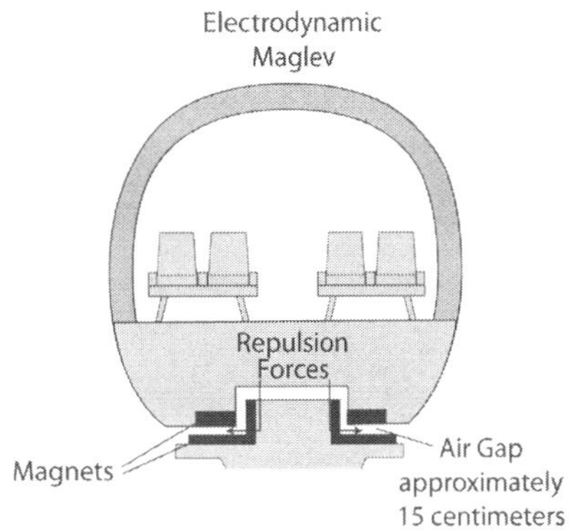

Figure 2

Researchers have compared the energy consumption of an electromagnetic suspension train (EMS) with a traditional high speed rail train (HSR). These results are shown in Figure 3.

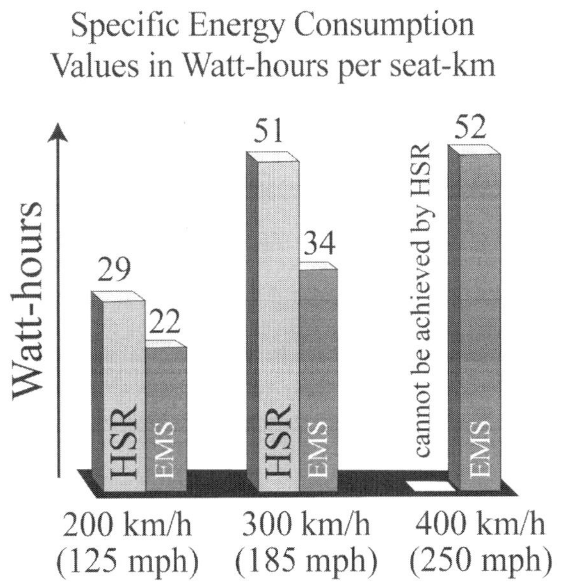

Figure 3

19. Describe the difference in separation between the rail and the train in the two suspension systems.

A The EDS air gap is 10x the EMS air gap.

B The EDS air gap is 5x the EMS air gap.

C The EMS air gap is 1/5 the EDS air gap.

D The EDS air gap is 1/10 the EMS air gap.

20. What is found in an EDS system that is never found in an EMS system?

F Electromagnets

G Permanent magnets

H Metal construction

J Repulsive field in the air gap

21. Using Figure 3, describe the energy consumption of the traditional train versus the EMS maglev.

A The traditional high-speed rail uses less energy than the EMS maglev.

B The EMS maglev uses more energy than the traditional high-speed rail train.

C The EMS maglev uses less energy than the traditional high-speed rail train.

D The two types of trains use almost the same amount of energy.

22. Which type of train requires the greatest energy increase to accelerate from 200 km/hr to 300 km/hr?

F The EMS maglev by 7 watt-hours

G The traditional high speed rail by 17 watt-hours

H The EMS maglev by 12 watt-hours

J The traditional high speed rail by 22 watt-hours

23. What is the top speed of the traditional high-speed rail train tested in Figure 3?

A <200 km/hr

B 20–300 km/hr

C 300–400 km/hr

D >400 km/hr

24. Which of the following is required for a maglev train to run?

F Gasoline

G Electricity

H Coal

J Air

Passage V

Ions are atoms that have gained or lost electrons. *Radius* is a measure of the size of the atom or ion, in picometers (10^{-12} meters). The radius of an ion helps us understand the effect of its electrical charge on other atoms and ions. Scientists have realized that the radius of an ion changes depending on its environment. The presence of a polar molecule, like water, can affect the size of an ion's radii. Polar molecules have a partial positive δ+ side (the H atom) and a partial negative δ– side (the O atom). The electrical nature of polar molecules can help to dissolve many solids. This can affect ionic radii in a measurable way.

Scientists carried out an experiment where they measured the difference in ionic radii of ions in hydrated form (dissolved in water), and unhydrated form (without water). They used various ions from three different rows (periods) on the Periodic Table. The higher the period number, the larger the atom is.

Their experimental results are shown in Tables 1 and 2 and are graphed in Figure 1. In Figure 1, hydrated values are labeled **H** and unhydrated are labeled **UN**.

Table 1

Period	Ion (Alkali metals)	Hydrated Radius (pm)	Ion (Alkali Earth Metals)	Hydrated Radius (pm)	Ion (Halogen)	Hydrated Radius (pm)
2	Li^{+}	382	Be^{2+}	459	F^{-}	352
3	Na^{+}	358	Mg^{2+}	428	Cl^{-}	332
4	K^{+}	331	Ca^{2+}	412	Br^{-}	330

Table 2

Period	Ion (Alkali metals)	Unhydrated Radius (pm)	Ion (Alkali Earth Metals)	Unhydrated Radius (pm)	Ion (Halogen)	Unhydrated Radius (pm)
2	Li^{+}	94	Be^{2+}	31	F^{-}	116
3	Na^{+}	117	Mg^{2+}	72	Cl^{-}	164
4	K^{+}	149	Ca^{2+}	100	Br^{-}	180

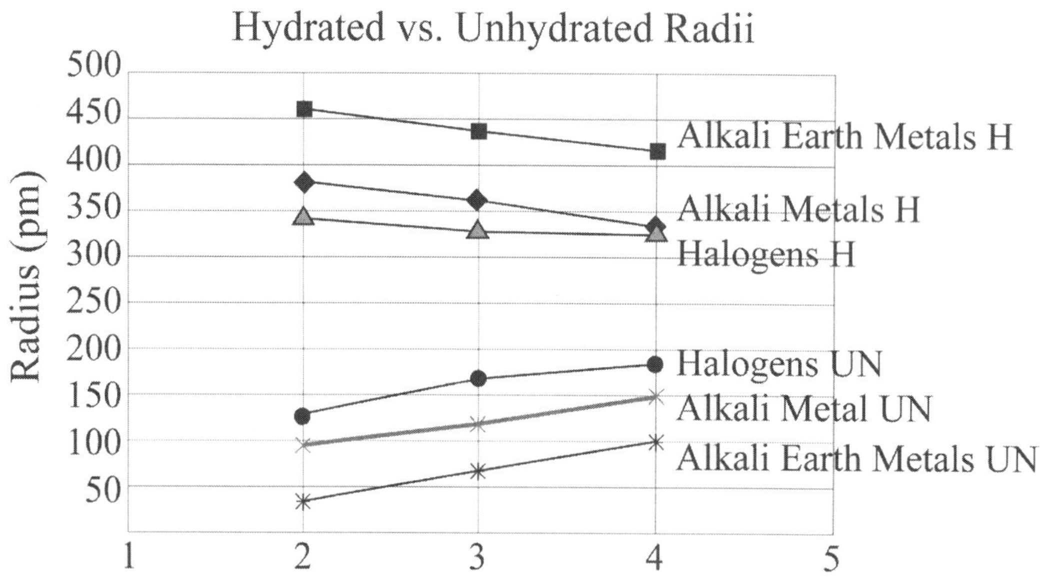
Hydrated vs. Unhydrated Radii
Radius (pm)
500
450
400
350
300
250
200
150
100
50
1
2
3
4
5
Alkali Earth Metals H
Alkali Metals H
Halogens H
Halogens UN
Alkali Metal UN
Alkali Earth Metals UN

Figure 1

25. From Table 2, the unhydrated radius of a calcium ion (Ca^{2+}) is approximately:

A the same as that of a potassium cation (K^{+}).

B half that of a chlorine anion (Cl^{-}).

C twice that of a magnesium cation (Mg^{2+}).

D three times that of a beryllium cation (Be^{2+}).

26. Based on Table 1, what is the hydrated radius of a sodium cation (Na^{+})?

F 3.58 10^{-12} m

G 3.58 10^{-10} m

H 1.17 10^{-12} m

J 1.17 10^{-10} m

27. Based on Figure 1, which group shows the largest difference between unhydrated and hydrated radii?

A Alkali metals

B Alkali Earth metals

C Halogens

D All groups show approximately the same difference.

28. According to Table 1 and Figure 1, the hydrated radii of fourth period halogens is MOST similar to:

F fourth period alkali metals.

G third period halogens.

H second period halogens.

J fourth period alkali Earth metals.

29. The element rubidium (Rb) is an Alkali metal found in Period 5 on the Periodic Table. Using Table 2, which of the following will best describe the unhydrated radius of the cation Rb^{+}?

A Greater than 149 pm

B Less than 149 pm

C Greater than 331 pm

D Less than 331 pm

Passage VI

A moving object's *momentum* is its tendency to remain in motion. For example, a large truck is harder to stop than a small car moving at the same speed. The truck therefore has more momentum than the car. Momentum can be defined as "mass in motion." An object's momentum is therefore dependent on the variables, mass and velocity. In equations terms, momentum is the product of mass and velocity, and it is reported in units of
kg × m/s.

When two objects collide with each other, they may bounce off each other or stick together. If two colliding objects rebound off each other at the exact speed at which they hit, the collision is said to be *elastic*. If they stick together, on the other hand, the collision is said to be *inelastic*.

Several students designed an experiment to test what happens to the momentum of a system during a collision. They used an air track with gliding carts to minimize contact between the cart wheels and the track. They used speed sensors. The experimental setup is shown in Figure 1. The students varied the amount of mass on Cart A. They accelerated Cart A to a velocity of 10 m/s. They then allowed Cart A to strike Cart B, which had been stationary. The two carts had magnets on their sides that caused them to stick together once they collided. The students then used a speed sensor to detect the velocity of Cart A and Cart B as they traveled together following the collision. Their results are shown in Table 1.

Before Collision

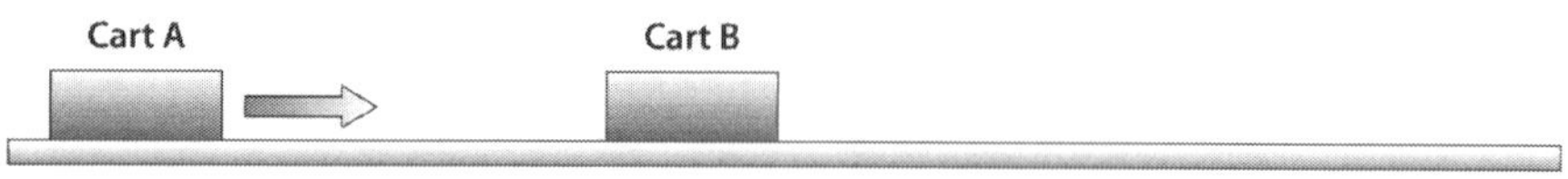

Figure 1

After Collision

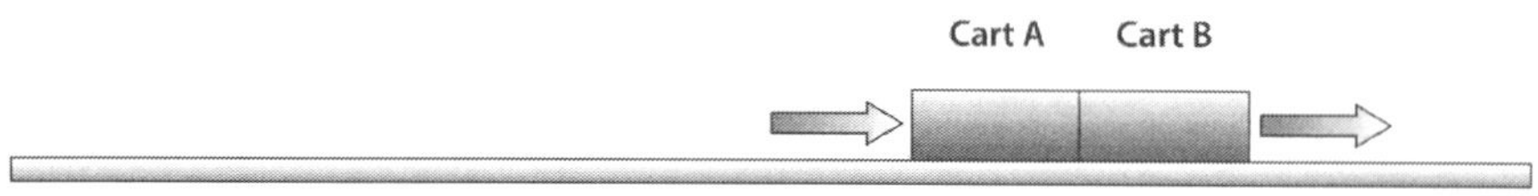

Figure 2

Table 1				
Results from Four Trials with Carts of Different Masses.				
Trial	Mass of Cart A (kg)	Mass of Cart B (kg)	Initial Velocity of Cart A (m/s)	Final Velocity of Carts A and B (m/s)
1	1	1	10	5
2	2	1	10	6.7
3	3	1	10	7.5
4	4	1	10	8

30. During the gliding cart experiment, was the collision of the carts elastic or inelastic?

F Elastic because they stuck together.

G Inelastic because they stuck together.

H Elastic because they bounced off each other.

J Inelastic because they bounced off each other.

31. Friction is the force that acts to slow down the relative motion of objects in contact. Friction between a rolling ball and the street, for example, will cause the ball to eventually slow to a stop. How did the experimenters try to limit the role of friction in the experiment?

A They ensured the initial speed of Cart A was consistent.

B They placed magnets between the carts so they would stick together.

C They varied the mass of Cart A.

D They used an air track with gliding carts.

32. What was the initial momentum of Cart A in Trial 1?

F 10 kg × m/s

G 5 kg × m/s

H 17 kg × m/s

J 1 kg × m/s

33. How does the change of the mass of Cart A affect the velocity of Carts A and B after collision?

A Increasing the mass of Cart A decreases the velocity of Carts A and B.

B Increasing the mass of Cart A increases the velocity of Carts A and B.

C Increasing the mass of Cart A does not change the velocity of Carts A and B.

D Increasing the mass of Cart A by 1kg increases the velocity of Carts A and B by 20%.

34. What was the final momentum of Carts A and B once stuck together in Trial 1?

F 5 kg m/s

G 10 kg m/s

H 20 kg m/s

J 2 kg m/s

35. Using the data in Table 1, which of the following conclusions can be drawn from this experiment?

A Decreasing the mass of an object will increase its velocity following a collision.

B Increasing the mass of an object will decrease its velocity following a collision.

C Increasing the mass of an object will increase its velocity following a collision.

D Decreasing the mass of an object will keep its velocity the same following a collision.

Passage VII

The Greenland ice sheet is the second largest body of ice in the world. This large ice mass is closely monitored by climate scientists. Melting of the ice sheet could cause a rise in sea levels and change ocean circulations, drastically altering the world's climate.

Scientists monitor the elevation at different areas of the ice sheet to estimate changes in the volume of the ice sheet overall. Changes in elevation are determined by accumulation and ablation. *Accumulation* is any process that adds snow or ice to an ice sheet, such as precipitation, condensation and refreezing of liquid water runoff. *Ablation* is any process that removes snow or ice from an ice sheet. Ablation is similar to runoff, but also happens through evaporation, sublimation and ice calving (when chunks of ice break and separate).

A scientist measured the elevation in the southwest and northeast margins of the ice sheet for the year 2001. During this year, these areas of the ice sheet were in balance with no significant changes in ice volume. He then measured the change in elevation from 2001 for the years 2002 through 2009 in the same regions of the ice sheet. He used annual accumulation and ablation estimates to determine the change in elevation from 2001 for each year. Elevation changes for the southwest and northeast margins are shown in Table 1. Elevation changes were measured in centimeters (cm).

Table 1

Year	Southwest Margin	Northeast Margin
2002	+11.2	+3.2
2003	–12.2	+15.6
2004	–26.7	+28.6
2005	–33.5	–15.4
2006	+12.9	–4.2
2007	+32.0	+23.5
2008	–6.8	+3.8
2009	–27.9	–19.6

36. What year served as the control year in this study?

F 2005

G 2009

H 2001

J 2002

37. In 2008, how did the southwest and northeast margins of the ice sheet change?

A Both areas receded.

B Both areas grew.

C The southwest grew, and the northeast receded.

D The southwest receded, and the northeast grew.

38. Which years experienced the greatest melting in the northeast and southwest?

F 2005 and 2009

G 2004 and 2005

H 2002 and 2007

J All years except 2002 and 2007

39. What weather parameter most likely contributed to changes in the ice sheet in 2007?

A High pressure

B Precipitation

C Evaporation

D Sublimation

40. Which of the following graphs most accurately reflects the data in Table 1?

F
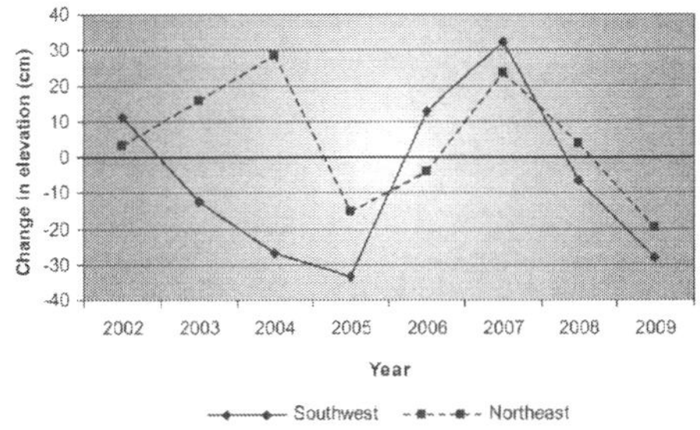

G
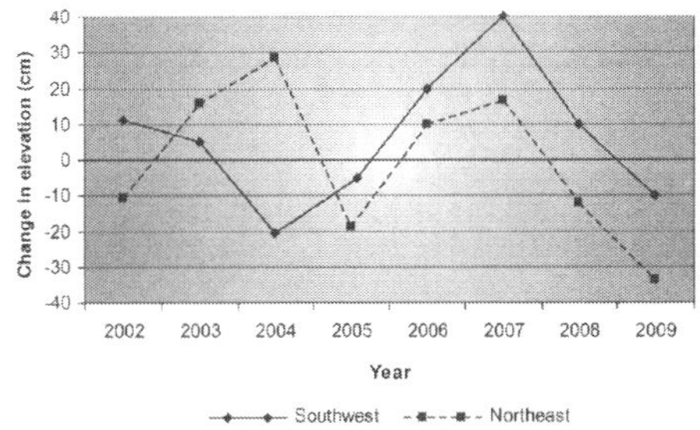

H
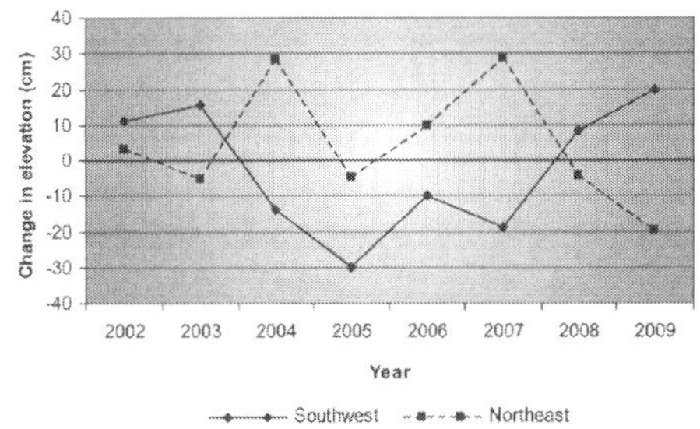

J
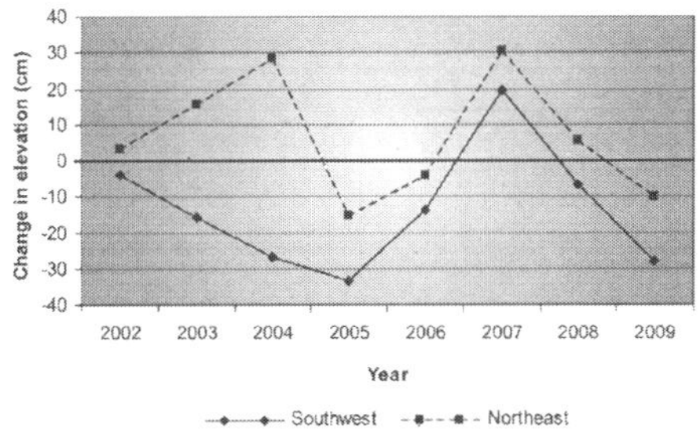

ACT Science Reasoning Test Evaluation Chart

Directions: Use the chart below to better focus your study for the ACT Science Reasoning Test. Identify the question range or questions you answered incorrectly. Turn to the appropriate chapters, read the explanations and complete the extra exercises. Review other chapters as needed. Finally, complete the practice tests as final preparation for the ACT Science Reasoning Test.

Chapter Number	Question Number
Chapter 2: Data Representation	14–18; 25–29; 36–40
Chapter 3: Research Summary	8–13; 19–24; 30–35
Chapter 4: Conflicting Viewpoints	1–7

Chapter 1
Preparing for the ACT Science Reasoning Test

The ACT Science Reasoning Test

This book focuses on the **Science Reasoning Test** portion of the ACT. When you take the ACT, you will be given 35 minutes to answer 40 questions in the Science Reasoning section. The 40 questions will be divided among 7 passages. Each passage describes some topic from a general field of science. These fields include Earth and space science, physics, chemistry and biology.

Right off the bat, it is important to emphasize that this ACT is not a "Science Test." It is a science *reasoning* test. The ACT Science Reasoning Test does NOT test your factual knowledge of science. All of the information that you need to answer the questions are contained in the passage itself. It tests your ability to obtain meaning from material on science topics. It requires you to think analytically. It does this in two ways: through reading comprehension and data analysis.

Reading Comprehension

The ACT Science Reasoning Test expects the student to read a passage and synthesize knowledge from it. That means that the student must be able to analyze the passage and interpret its meaning.

Data Analysis

The test also requires the student to think symbolically. That means that you must be able to describe the trends shown by the data contained in tables and graphs. You must also describe the parts and purpose of the elements of a diagram.

These two skills are the core of the ACT Science Reasoning Test. One or both of these skills are required to answer the questions that accompany every type of passage on the ACT Science Reasoning Test. The types of passages that are represented on the test are: **Data Representation (DR), Research Summaries (RS) and Conflicting Viewpoints (CV).**

The chapters in this book focus on reviewing the reading comprehension and data analysis skills that you will need to answer each type of question on the test.

The chapters do not teach science content, which you have spent time learning throughout your education. Instead, they offer examples and help you to practice the specific types of questions that you will encounter on the ACT. The questions used in the exercises throughout the chapters and in the review at the end of each chapter are very similar in wording and focus to actual questions you will see on the ACT Science Reasoning Test.

The diagnostic test at the beginning of this book and two practice tests at the end of the book are simulated ACT tests. They are the same length and contain passages and questions comparable to those you will see on the ACT. Review your scores on these practice tests with your teacher or tutor to determine if there are skill areas you need to hone before taking the actual ACT. Time yourself while taking these simulated tests to correctly adjust your pace during the actual ACT exam.

For practice with other sections of the ACT, refer to these titles from American Book Company:

- *ACT Mathematics Test Preparation Guide*
- *ACT Reading Test Preparation Guide*
- *ACT English Test Preparation Guide*

Frequently Asked Questions

What Is the ACT?

The ACT is a national college admission examination which is accepted by virtually all United States colleges and universities. The exam measures the knowledge, understanding and skills that you have gained throughout your education.

How Long Is the Exam?

The full ACT includes 215 multiple-choice questions and takes approximately three hours and thirty minutes to complete (just over four hours if you are taking the optional Writing Test). The exam has four sections: English, math, reading and science.

When Do I Take the ACT?

The ACT is administered in October, December, February, April and June (in some states, it is also offered in September). When you take the ACT will depend on both when you feel ready and when you need test scores to include with applications to colleges.

Is There a Fee?

Yes. The basic registration fee (currently $30.00 for the basic test or $44.50 if also taking the Writing Test) includes score reports for up to four colleges for which a valid code is listed at the time that you register.

For more details about the ACT, please visit ACT, Inc. at www.actstudent.org.

Tips for ACT Preparation and Testing

The ACT measures your overall learning, so if you have paid attention in school, you should do well! It would be difficult, if not impossible, to "cram" for an exam as comprehensive as this. However, you can study wisely by using an ACT-specific guide (like this book) and practice answering questions of the type that will be asked on the ACT (included in this book; additional practice tests are available online from ACT, Inc.).

Preparing for the ACT

- Believe in yourself! Attitude plays a big part in how well you do in anything. Keep your thoughts positive. Tell yourself you will do well on the exam.
- Be prepared. Get a good night's sleep the day before your exam. Eat a well-balanced meal, one that contains plenty of proteins and carbohydrates, prior to your exam.
- Arrive early. Allow yourself at least 15–20 minutes to find your room and get settled. Then, you can relax before the exam, so you won't feel rushed.
- Practice relaxation techniques. Some students become overly worried about exams. Before or during the test, they may perspire heavily, experience an upset stomach or have shortness of breath. If you feel any of these symptoms, talk to a close friend or see a counselor. They will suggest ways to deal with test anxiety. Here are some quick ways to relieve test anxiety:
    - Imagine yourself in your favorite place. Let yourself sit there and relax.
    - Do a body scan. Tense and relax each part of your body starting with your toes and ending with your forehead.
    - Use the 3-12-6 method of relaxation when you feel stress. Inhale slowly for 3 seconds. Hold your breath for 12 seconds, and then exhale slowly for 6 seconds.

Taking the ACT

- Carefully **read the instructions** on the ACT test booklet. You can read these instructions ahead of time at www.actstudent.org to make sure that you understand them.
- Once you are told that you may open the booklet, **read the directions for each test** thoroughly before beginning to mark answers.
- **Read each question carefully**, and use your best approach for answering the questions. Some test-takers like to skim the questions and answers before reading the problem or passage. Others prefer to work the problem or read the passage before looking at the answers. Decide which approach works best for you.
- **Answer each question** on the exam, as your score is based on the number of questions answered correctly. There is no penalty for guessing, but every spot left blank is automatically a zero.

- If you are uncertain about an answer, **take an educated guess**. Eliminate choices that are definitely wrong, and then choose from the remaining answers.

- **Use your answer sheet correctly**. Make sure the number on your question matches the number on your answer sheet. If you need to change your answer, erase it completely. Use a number two pencil, and make sure the answers are dark. The computerized scanner may skip over answers that are too light.
- **Check your answers**. If you finish a test before time is called, review your exam to make sure you have chosen the best responses. Change answers only if you are sure they are wrong.
- Be sure to **pace yourself**. Since you will have a limited amount of time, be careful not to spend too much time on one passage or problem, leaving no time to complete the rest of the test. Listen for the announcement of five minutes remaining on each test.
- When time is called for each test, **put your pencil down**. If you continue to write or erase, you run the risk of being dismissed and your test being disqualified from scoring.

When Taking the Science Reasoning Test

As we mentioned at the beginning of this chapter, there are three different kinds of passages on the ACT Science Reasoning Test. Now, we will look at them in a little more detail.

Data Representation (DR)

DR passages are designed to show sets of data. A brief passage that is accompanied by tables, charts, figures and/or graphs will be given. You are to interpret this data. There are 3 data representation passages per test, each with 5 questions. These 15 questions make up 38% of the ACT Science Reasoning Test.

Research Summaries (RS)

RS passages require you to read and understand two or more experiments and their results. You are tested on whether you understand what the scientists accomplished and their results. There are 3 research summary passages per test, each with 6 questions. These 18 questions make up 45% of the ACT Science Reasoning Test.

Conflicting Viewpoints (CV)

CV passages present the research and conclusions of two or more individuals that have reached different conclusions about a science topic. You are required to describe the source of the conflict and the evidence that supports each person's conclusion. There is 1 conflicting viewpoint passage per test with 7 questions. These 7 questions make up 17% of the ACT Science Reasoning Test.

Remember that you will not be tested on your recall of science facts. You will be tested on your understanding of passages describing science topics and your ability to determine the meaning behind the information and data presented in those passages.

Chapter 2
Data Representation

DR passages have 5 questions

Data is factual information. Data can come from anywhere. For instance, you could record the grade that you earned for every class you took in your junior year of high school. That is data. You can use data to analyze problems. If you look at your class by class grade list, you could probably see what subjects you excel in and what subjects you need to work harder in. Scientists generate data from scientific experiments. They use the data to draw conclusions about scientific questions.

Experimental data must be presented in an organized manner. Data presented in an organized manner is easier to analyze and interpret. By carefully evaluating the data, you can interpret the results of the experiment. Tables, graphs, models and diagrams are all ways of presenting the results of an experiment.

TABLES

The most common type of data presentation on the ACT Science Reasoning Test is the table. A **table** is a good way to organize data. It presents information in an orderly fashion, usually in rows and columns.

There are as many table formats as there are data types. A table can list characteristics of items. For example, Table 2.1 contains the names and colors of five different species of rose.

Table 2.1

Rose	Color
Sulfur rose	yellow
Maiden's Blush	pale pink
Lady Banks' rose	peach
Oklahoma	red
Cherokee	white

Table 2.1 is a **qualitative** data table. It is basically an orderly list of non-numerical things. More commonly, however, you will see the **quantitative** data table. This is an orderly listing of numerical items.

To show an example of this kind of table, let's expand Table 2.1. Say a researcher is counting the number of bushes of each rose species found in a particular area. A scientist that researches plants (a botanist) would certainly want to use the scientific name of the flower. Table 2.2 shows how we enlarged Table 2.1 to include this information.

Table 2.2

Common Name	Scientific Name	Color	Number of Bushes
Sulfur rose	*R. hemisphaerica*	yellow	18
Maiden's Blush	*R. alba*	pale pink	47
Lady Banks' rose	*R. banksiae*	peach	3
Oklahoma	*R. odorata*	red	9
Cherokee	*R. laevigata*	white	26

The ACT Science Reasoning Test is filled with data like this. You must be comfortable with analyzing data in order to succeed on this test. The more data that you see, the more comfortable you will be, so let's move on. Through the next several sections, we will be using the data in Table 2.1 again and again. We will build on this data and manipulate it to show several formats for presentation. Along the way, we will stop and ask specific questions about each data format.

Analyzing a Data Table Part 1

The most basic part of analyzing a data table is locating information within the table. The **columns** of a table are the vertical categories. There are four columns in Table 2.2. The columns are labeled Common Name, Scientific Name, Color and Number of Bushes. Column headings describe the *characteristics of a thing.*

Rows are the horizontal part of a table. There are five rows in Table 2.2. The rows begin with the row headings: Sulfur rose, Maiden's blush, Lady Banks' rose, Oklahoma and Cherokee. They are all common names for different types of roses. If you let your eye slide to the right after the row header "Cherokee," you see that the Cherokee rose has a scientific name of *R. laevigata*, that it is white and that 26 bushes were found in the area that the botanist studied. So, you see, the row heading contains *the thing itself* that is being described by the characteristics found in the columns.

Practice Questions 1

1. Which rose species has the common name Maiden's blush?

A *R. hemisphaerica*
B *R. alba*
C *R. banksiae*
D *R. laevigata*

The best answer is B. This answer can be obtained by looking directly at the column containing the common name of the rose bushes. Locate the common name Maiden's Blush, and then move across the row to find the scientific name. Choice A (*R. Hemisphaerica*) has the common name of Sulfur rose. Choice C (*R. banksiae*) has the common name of Lady Banks' rose. Lastly, Choice D (*R. laevigata*) is the common name of Cherokee.

2. The botanist found 9 bushes of a particular kind of rose. What rose was that?

   A *R. hemisphaerica*

   B *R. banksiae*

   C *R. odorata*

   D R. laevigata

3. Which of the following roses are yellow?

   A Sulfur rose

   B Maiden's Blush

   C Oklahoma

   D Cherokee

Analyzing a Data Table Part 2

Tables are also used to show data that changes with time. Let's look back at our botanist. Now she is counting the number of bushes for each rose species found in her particular area of study in different months of the year.

Table 2.3				
Scientific Name of Species	Jan	April	July	Oct
R. hemisphaerica	18	18	18	18
R. alba	47	47	47	40
R. banksiae	3	3	3	0
R. odorata	9	9	9	7
R. laevigata	26	26	27	27

Here we can see the changing numbers of different species of rose, between January and October. There are two things that you should be paying attention to at this point:

1. **Examine the data to determine trends in each species.** How is the population of *each species* changing with time? To figure this out, look at the numbers in each *row*. Example: The population of *R. hemisphaerica* does not change at all over the study period. How do you know this? The number of *R. hemisphaerica* bushes found is 18 in every month of the study period.

2. **Examine the data to determine trends in *all* species.** How is the population of *all species* changing with time? To figure this out, look at the numbers in each *column*. Here's an example: The total rose population is exactly the same in January and April. How do you know this? Compare the January and April columns. The numbers of each rose species are exactly the same in those two months. Another way to figure this out is to actually add up the number of each rose species for each month of the study period.

Test Tip: Become very familiar with tables.

Practice Questions 2

1. Which species' population changes the most during the study period?

   A *R. hemisphaerica* **C** *R. banksiae*

   B R. alba **D** *R. laevigata*

The best answer is B. Careful examination of the range of the data reveals the answer. The range of the data is the difference between the highest and lowest number. To find the range subtract the lowest from the highest number. The answer is your range. This tells you which rose population changed the most during the study period. Choice B shows a range of seven plants (47 – 40 = 7), which is the greatest population change. Choice A does not experience a population change during the study period (18 – 18 = 0), so its range is zero. Choice C is reduced by three plants (3 – 0 = 3) during the study period, so its range is three. Choice D changed by a total of one plant (27 – 26 = 1) during the study period.

2. The botanist found that one rose species increased in numbers during the study period. What rose was that?

   A *R. hemisphaerica* **C** *R. odorata*

   B *R. banksiae* **D** R. laevigata

3. Which species of rose was no longer growing in the area of study, at the end of the study period?

   A *R. hemisphaerica* **C** *R. banksiae*

   B *R. alba* **D** *R. laevigata*

From Data Table to Graph

A table is a great way to organize data, and it allows you to see trends in data. But, there is another way to see trends: graphs. **Graphs** are a visual way to represent data. To correctly design a graph, you must recognize the variables presented in a data table. There are two types of variables.

The **independent variable** is the thing that you are testing. In Table 2.2, the independent variable is the scientific name of the rose. Can you guess the independent variable found in Table 2.3? Right! The months in the study period.

The **dependent variable** is the *change* in the thing that you are testing. It is the data collected. In our data set, the dependent variable is the change in the number of roses, either by species or per month. For both Table 2.2 and 2.3, the dependent variable is the number of rose bushes found in a particular area.

Let's look at the three most common types of graphs. We will see how they fit our data.

THE BAR GRAPH

A **bar graph** is a very common type of graph. It is nearly always in the ACT, often several times per test. The bar graph is usually used to compare the numbers of different things. It uses rectangular bars to show how large each value is and can either be horizontal (most common) or vertical. Figure 2.1 shows the general format of a bar graph.

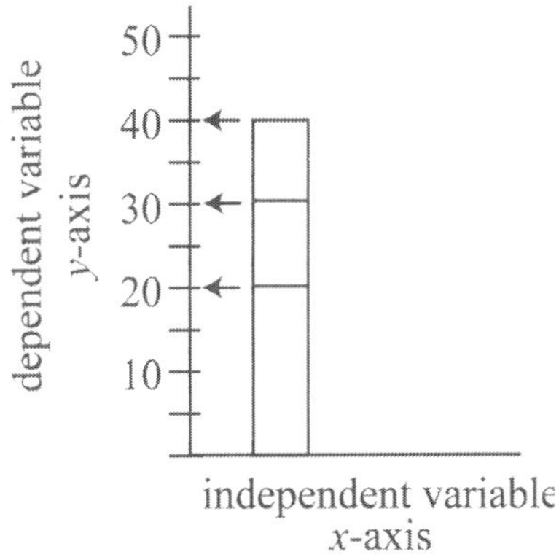

Figure 2.1

The independent variable is graphed on the horizontal axis. The horizontal axis is called the ***x*-axis**. In a bar graph, the *x*-axis usually contains qualitative information, like rose species. A time segment, like a month, may also be placed on the *x*-axis.

The dependent variable is graphed on the vertical axis. The vertical axis is called the ***y*-axis**. The *y*-axis contains quantitative information—that is, it contains the number or data you are trying to collect during the experiment.

The height of the bar shows the number of the independent variable. The higher the bar, the greater the number. The number of the independent variable can be determined by looking at the *y*-axis. Let's use some bar graphs of our rose data to fully explain the use of the bar graph. These are shown in Figures 2.2 and 2.3.

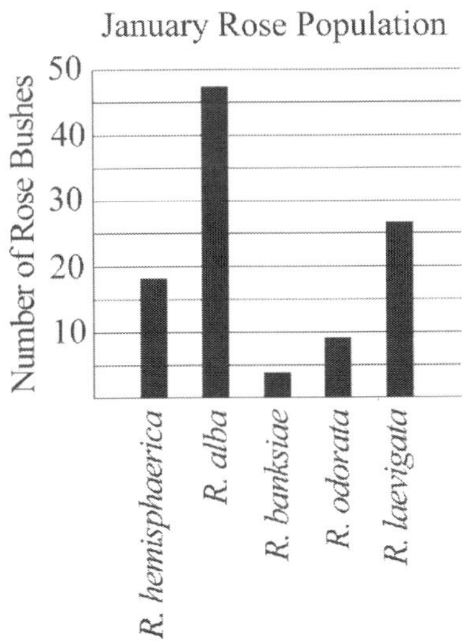

Figure 2.2

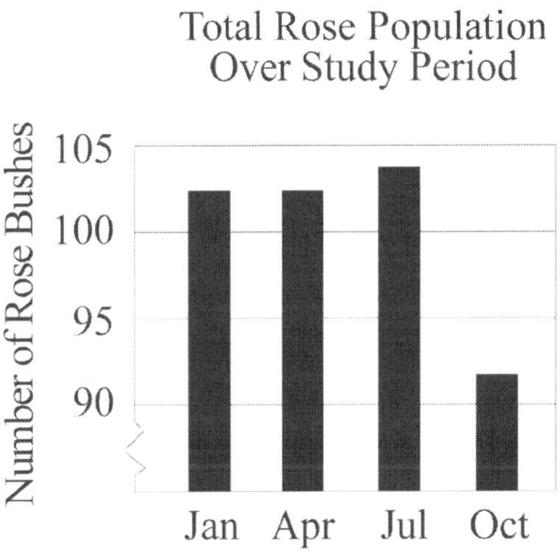

Figure 2.3

Here are three important things to quickly do when you see a bar graph:

1. **Examine the variables on the axes.** These two bar graphs describe very different things about the rose data. Figure 2.2 describes the variation in population of each rose population in the month of January. Figure 2.3 describes the variation in total rose population throughout the study period.

2. **Examine the range of the axes.** In Figure 2.2, the axis begins at zero and goes up to 50; the range is 50. Figure 2.3 has a different kind of axis, though: it does not start at zero. The two lines at the base of the *y*-axis shown are called a **break**. They show that the numbering of the axis is starting at a number higher than zero. If the break had not been put in, the graph would have been really tall, and we would not be able to see the differences between the bars very well. The range of the *y*-axis in Figure 2.3 is 105 – 90, or 15. While the ACT will not ask you to calculate the range, it will ask you to determine the value associated with a data point. To do that, you need to understand both the range of the axes and their scale.

3. **Examine the scale of the axes.** The **scale** is the amount associated with each interval or tick mark on the axis. In Figure 2.2, each labeled interval equals 10 rose plants. In Figure 2.3, each labeled interval equals 5 rose plants.

The ACT will ask you both qualitative and quantitative questions about bar graphs. Qualitative questions will ask you to describe data with words. Quantitative questions will ask you to describe data with numbers. This means reading numbers off the graph. It also means that you should be able to describe changes in data with percentages or fractions.

> ***Test Tip:*** In most cases, the data that the bar graph is based on will not be shown. Only the bar graph will be shown. You will have to interpret the graph to obtain the data.

Practice Questions 3

1. Which species had the lowest population in January? (Qualitative Version)

   A *R. odorata*   **C** *R. banksiae*

   B R. alba   **D** R. laevigata

The best answer is C. Figure 2.2 shows the rose population in January by species. Use the graph to answer the question. Choice C has the smallest bar seen in the graph. Choices A, B and D all have larger bars and consequently higher populations.

2. What was the population of *R. banksiae* in January? (Quantitative Version)

   A 2   **C** 4

   B 3   **D** 5

3. Describe the change in overall rose population over the study period. (Qualitative Version)

   A It decreased each month between January and October.

   B It increased slightly during the study period.

   C It stayed steady, then dropped sharply in October.

   D It stayed steady throughout the study period.

4. Describe the change in overall rose population over the study period. (Quantitative Version)

   A It decreased by 1/2 between July and October.

   B It increased by 1/4 between January and April.

   C It decreased by about 10% between July and October.

   D It decreased by about 20% between July and October.

The Line Graph

A **line graph** is another type of graph commonly found on the ACT. The line graph is used to show trends in data. A **trend** is an expression of how things change, usually as related to time. Figure 2.4 shows the general format of a line graph.

Line graph questions are usually about the trends shown, rather than a quantitative assessment of specific numerical data.

As with bar graphs, you should examine the axes to determine the variables being tested, as well as the range and scale of the axes. Note that line graphs have a scale on *both* axes; the scale on the *x*-axis may well be different from that shown on the *y*-axis. In addition, you should know these important facts:

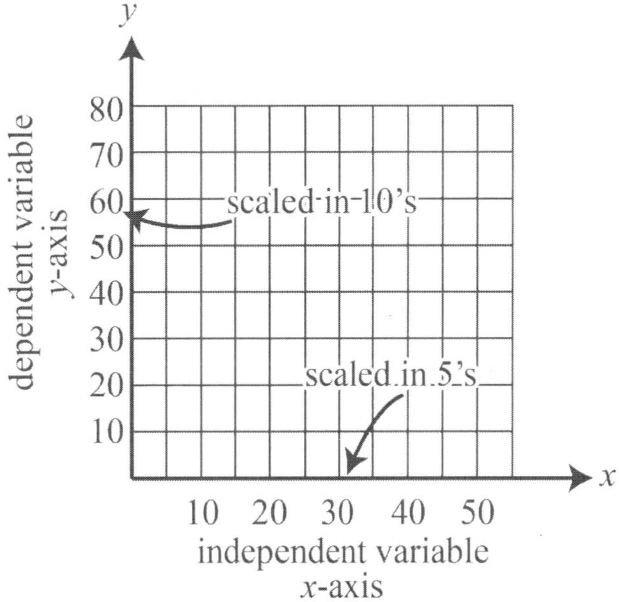

Figure 2.4

1. **Know how to analyze the trendline.** A **trendline** is the line that connects all the plotted data. When the value of the variable plotted on the *y*-axis is increasing, it is called an **increasing trend**. When the value of the variable plotted on the *y*-axis is decreasing, it is called a **decreasing trend**. When the value of the variable plotted on the *y*-axis is not changing, there is no trend; the system is at **equilibrium** (a state of zero net change). In such a case, the variables may probably be unrelated.

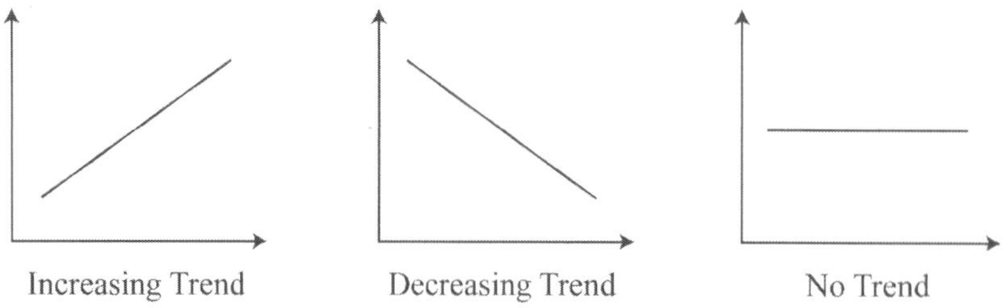

Figure 2.5

2. **Understand that a trendline is generated from individual data points.** The individual data points are then mathematically averaged into an equation. If the trendline is straight, its equation will fit the form $y = mx + b$. In this equation, m is the slope. The greater the value of m, the more positive the slope and the faster the trend is increasing. The more negative the slope, the faster the trend is decreasing. When the line is flat, the slope is zero, and the trend is neither increasing nor decreasing.

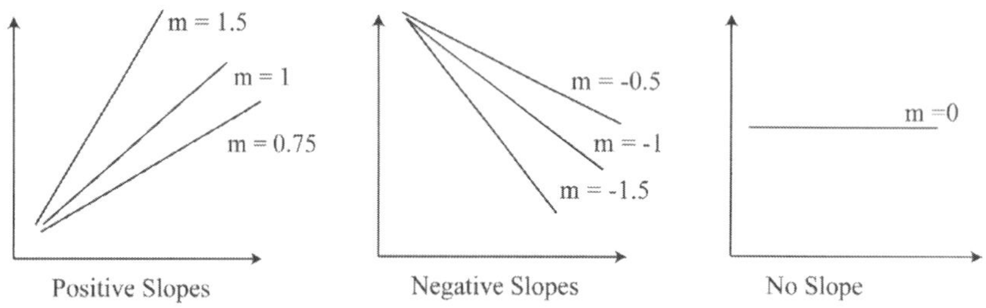

Figure 2.6

3. **Individual data points may also be shown.** Sometimes, the trendline is shown with the individual data points. Sometimes, only the data points are shown, without the trendline. This is called a **scatter plot**. With or without the trendline, showing the scatter plot lets you see how wide or clumped the data points are. This gives us a general idea of the usefulness of the trendline. Generally, highly clumped data points make poor trendlines. More spread out data usually create a more reliable or trustworthy trendline.

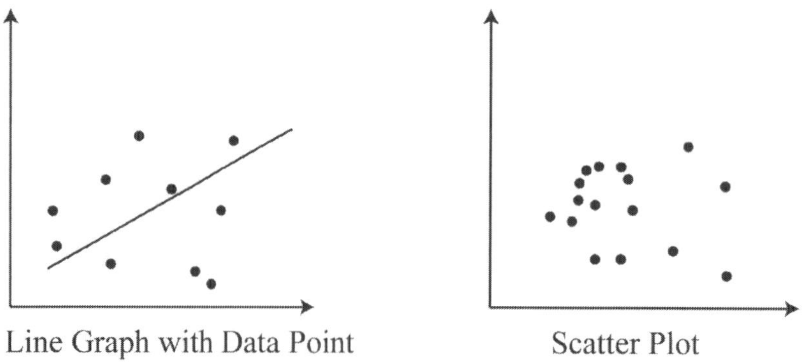

Figure 2.7

4. **The ACT often tests your ability to extrapolate. Extrapolation** is the estimation of future data from the trendline. For instance, our botanist took data from January to October. If she used her trendline to predict data in December, she would be extrapolating. To do this, simply use your pencil to continue the straight line of the trendline out to the *x*-axis point that you are hoping to estimate—in this case, 12 (or December). Using the extrapolated trendline, we can see that the *y*-axis point that corresponds directly to the month of December is fifteen rose plants.

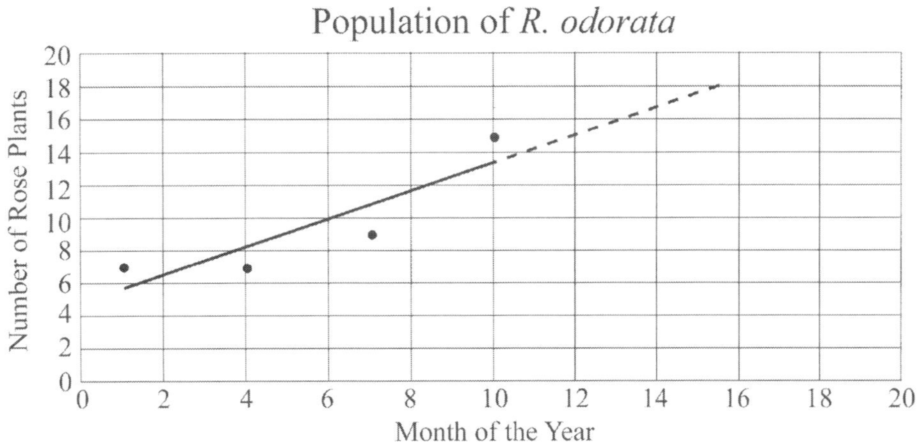

Figure 2.8 (a)

5. **Estimating a value that falls withing the range of known values. Intrapolation** is another type of graph problem that the ACT might test. In the case of our botanist, we can extimate data that falls between any two months in which data was collected and recorded. The botanist did this for the months of January, April, July and October. Using the trendline, she can intrapolate to estimate for all the months in between for which data was not collected. For instance, Figure 2.8b shows how estimates for February and March can be found.

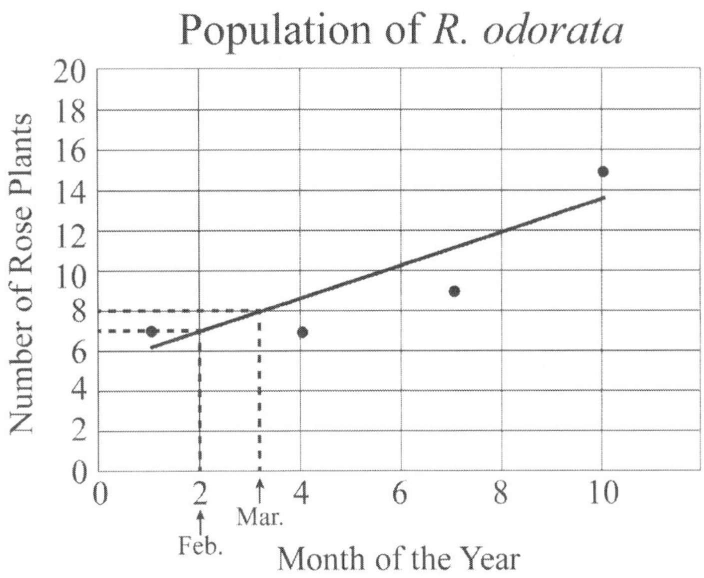

Figure 2.8 (b)

6. **Several trendlines may be shown on one graph.** This is called a **multiple line graph**. It allows you to examine the changes in several data sets at one time on the same scale.

Let's expand on our botanist's earlier efforts. Let's say that she decides to continue her study for another year. She makes measurements in the same months of January, April, July and October. Her results are shown in Figure 2.9.

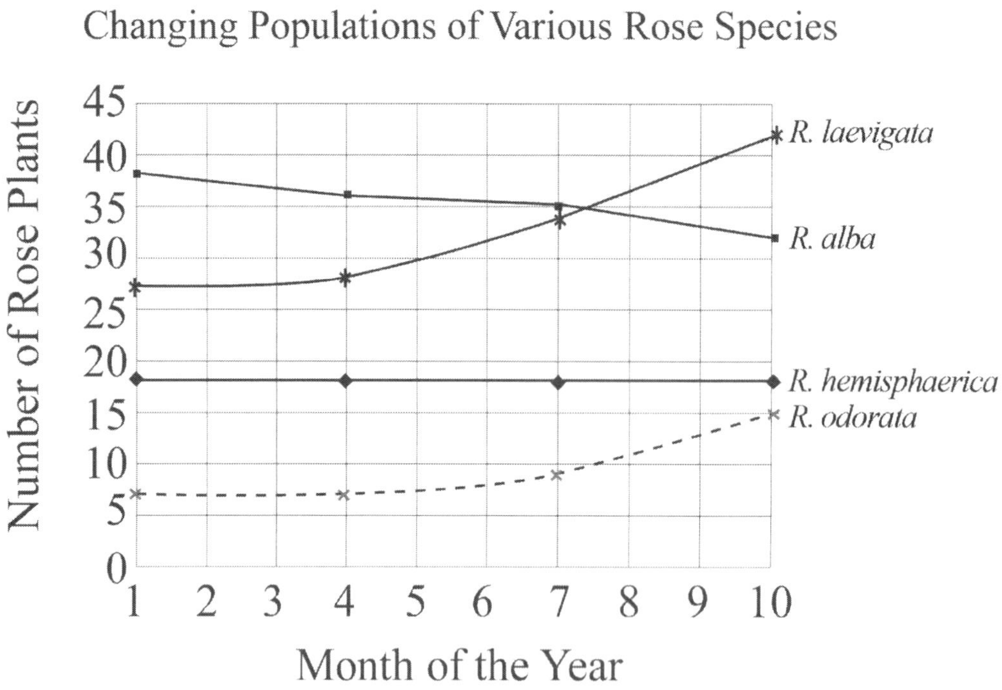

Figure 2.9

Practice Questions 4

1. The botanist finds that only one rose species shows a decreasing population in this second year of study. Which one is it?

A *R. hemisphaerica* **C** *R. odorata*

B *R. alba* **D** R. laevigata

The best answer is B. This is the only rose species with a negative trendline seen in the study. Choice A has zero population change. The trendline for both Choices C and D show an increasing trend.

2. Which species had the largest population in January?

A *R. hemisphaerica* **C** *R. odorata*

B *R. alba* **D** *R. laevigata*

3. Which species had the greatest population in October?

A *R. hemisphaerica* **C** *R. odorata*

B *R. alba* **D** *R. laevigata*

4. Based on the trendlines, which species' population is at equilibrium?

A *R. hemisphaerica* **C** *R. odorata*

B *R. alba* **D** *R. laevigata*

5. The botanist calculates a trendline for *R. odorata*. Based on the trendline, she makes a prediction about the population of that species in April of the following year. What would her prediction be?

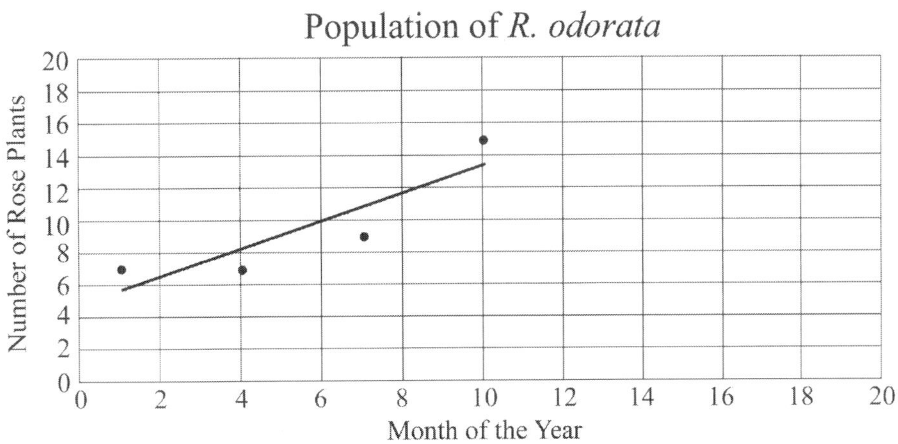

A 10 plants **C** 18 plants

B 12 plants **D** 22 plants

The Circle Graph

The circle graph (or a pie chart) is a third general type of graph. It is less commonly seen on the ACT, mainly because it has a very limited purpose. A **circle graph** is used to show the relative parts of a whole. It is excellent for visually estimating percentages. In order for you to quickly estimate the percentages shown on a circle graph, it is important to remember the following information:

- The full circle equals 100%.
- A half (1/2) circle equals 50%.
- A third (1/3) of the circle equals 33%.
- A quarter (1/4) of the circle equals 25%.

Rose Species in April

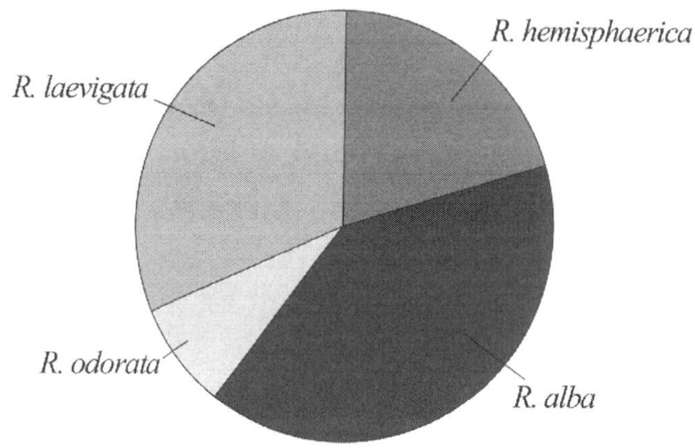

Practice Questions 5

1. *R. laevigata* represents approximately what percent of the rose population?

A 15% **B** 20% **C** 30% **D** 50%

The best answer is C. Slightly more than 25% or one quarter (¼) of the graph is occupied by *R. Laevigata*. Only Choice C correctly describes this situation. The population of *R. odorata* more accurately corresponds with Choice A. *R. hemisphaerica* more accurately corresponds with Choice B. Choice D is too large; no rose species occupies 50% of the population.

2. The population of one rose species is over 33%. Which one is it?

A *R. hemisphaerica* **C** *R. odorata*

B R. alba **D** *R. laevigata*

3. A horticultural journal indicates that rose species that represent less than 1/4 of the total rose population in a given area are most likely to become endangered. Which species of rose in the study area could become endangered?

   A *R. hemisphaerica* and *R. laevigata*

   B *R. alba* and *R. odorata*

   C *R. alba* and *R. laevigata*

   D *R. hemisphaerica* and *R. odorata*

SUMMARIZING DATA REPRESENTATION

Data Representation Review

DR passages have 5 questions

In order to answer DR questions, you must:

- Analyze and compare raw data
- Carefully review rows and columns in data tables for trends
- **Independent variable** is the topic under investigation (*x*-axis)
- **Dependent variable** is the data collected (*y*-axis)
- **Bar graphs** compare the amounts of different things. You must be able to:

  – examine axes

  – determine range and scale
- **Line graphs** show trends in scatter plot data. Sometimes there are multiple lines on one graph. You must be able to:

  – interpret trendlines

  – extrapolate future data points
- **Scatter plots** show individual data points. You must be able to:

  – determine distribution of data
- Circle graphs show parts of a whole. You must be able to:

  – estimate percentages or fractions

The types of data analysis that we have discussed are the most common types found on the ACT. There may be different diagrams that go along with the data and support it, but the main purpose of the Data Representation questions is to assess your ability to analyze and compare raw data.

Chapter 2 Skill Review

Passage I

Primary productivity is the amount of carbon that autotrophs can fix as sugars. This fixation usually occurs through the process of photosynthesis. Autotrophs require sunlight to make sugar by photosynthesis.

The reverse of this process is respiration, which all organisms use to release energy from sugars. Because photosynthesis and respiration both involve oxygen, dissolved oxygen is used as a tracer to measure primary productivity. Scientists measure dissolved oxygen by placing equal amounts of phytoplankton in pairs of bottles. One of the bottles in each pair is clear (light), and the other is opaque (dark). Pairs of bottles are sealed and lowered to different depths in the ocean. At each depth, both a clear and a dark bottle are present. The bottles are kept at constant depth in the same location for a set amount of time.

Oxygen levels change in the dark bottle because of respiration which consumes oxygen. Oxygen levels change in the clear bottle because of both respiration (consuming oxygen) and photosynthesis (releasing oxygen). The opaque bottle, in which no photosynthesis can occur, is an experimental control. Against it, we can compare the photosynthesis in the light bottle. After several hours, the bottles are pulled out of the water and their oxygen contents are measured. The measurements are given in Table 1.

Table 1

Depth	% Oxygen Change		Depth	% Oxygen Change	
	Clear bottle	Opaque bottle		Clear bottle	Opaque bottle
10 m	+ 12	–21	60 m	+3	–20
20 m	+ 19	–20	70 m	0	–21
30 m	+ 15	–21	80 m	–10	–20
40 m	+ 10	–19	90 m	–16	–19
50 m	+ 7	–20	100 m	–21	–20

1. At what depth is primary productivity greatest?

A 10 meters
B 20 meters
C 70 meters
D 100 meters

2. What is occurring in the clear bottle at 50 meters?

F Photosynthesis only
G Respiration only
H Both photosynthesis and respiration
J Neither photosynthesis nor respiration

3. Which of the following graphs best represents the data?

A

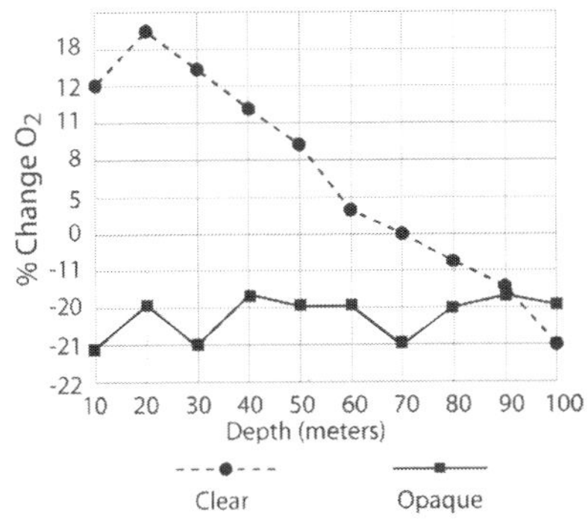

B

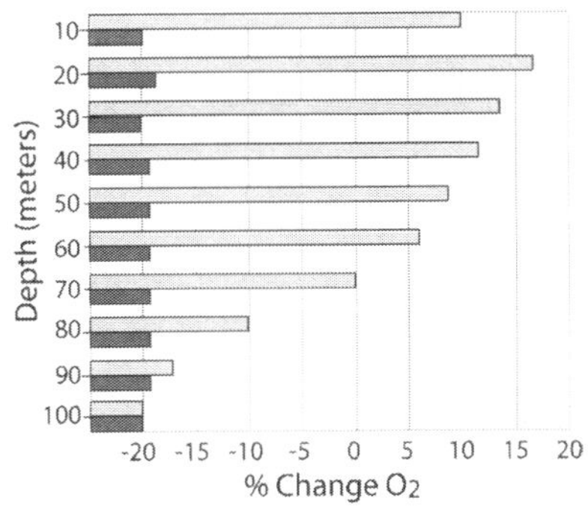

C

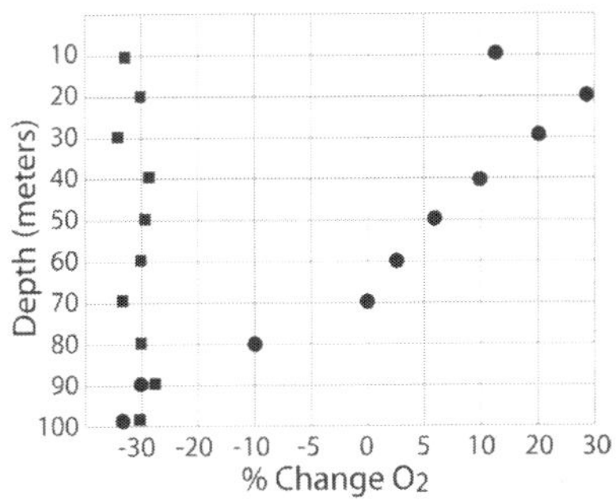

D

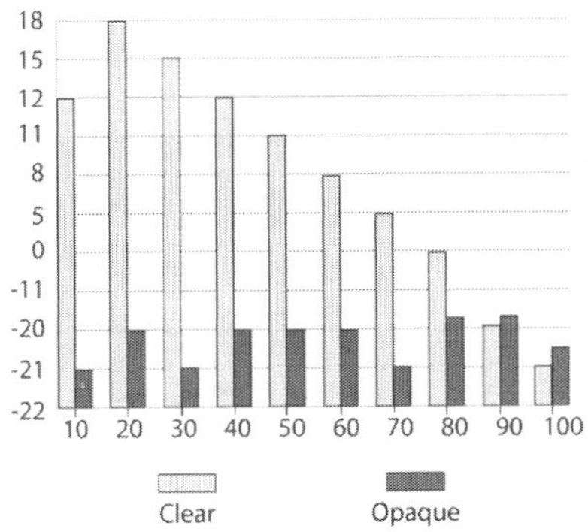

4. The compensation depth is the depth at which the oxygen change in the light bottle is zero. This occurs at 70 meters. What is the significance of the compensation depth for respiration and photosynthesis?

F The production and consumption of oxygen are equal.

G No oxygen is produced or consumed.

H Respiration rates are higher than photosynthetic rates.

J Photosynthetic rates are higher than respiration rates.

5. In what range of depth does photosynthesis stop?

A Between 20 and 40 meters

B Between 40 and 50 meters

C Between 60 and 80 meters

D Between 90 and 100 meters

Passage II

Recycling saves energy. Products made from non-recycled raw materials use more energy in the collection, transportation, processing and manufacturing of the materials. Therefore, making products with recycled material reduces this process and saves energy. The energy savings applies to all recycling sectors. An environmental organization estimated the percentage of several materials recycled in 2007 in the United States (Figure 1). They also estimated the amount of energy that would be saved by recycling 50%, 75% and 100% of each raw material (Figure 2).

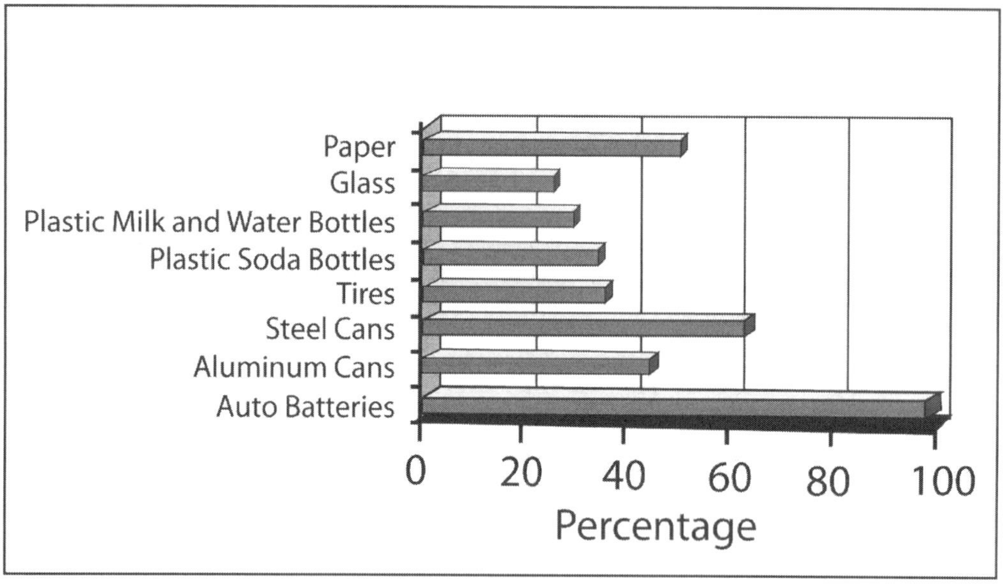

Figure 1

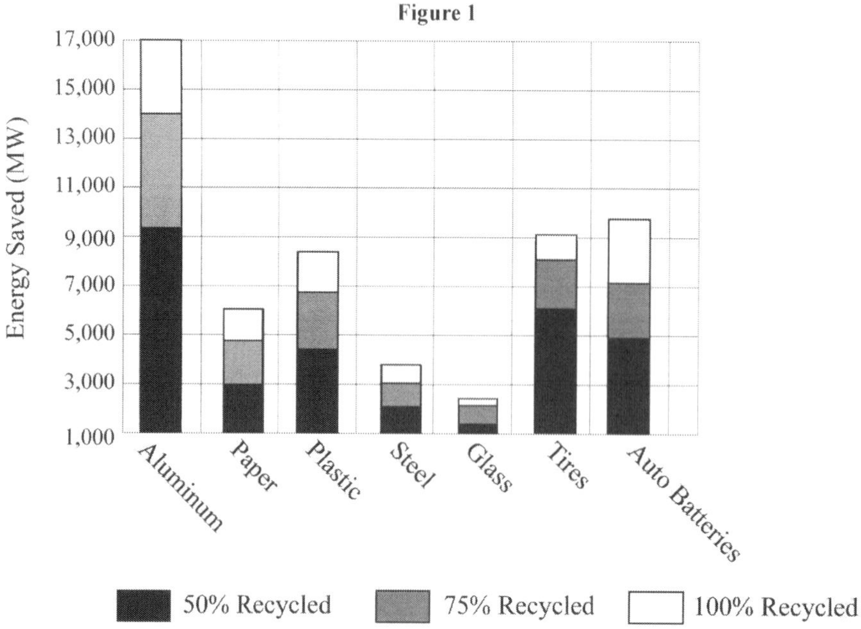

Figure 2

1. Milk and water bottles are made from a different type of plastic than soda bottles. Based on Figure 1, which type of bottle is most commonly recycled?

   A Milk bottles

   B Water bottles

   C Soda bottles

   D Milk and water bottles

2. How much energy could be saved if 75% of plastic was recycled?

   F About 5,000 MW

   G About 7,000 MW

   H Between 7,000 and 9,000 MW

   J About 9,000 MW

3. How much more energy would be saved if the recycling rate of steel rose from its current rate to 100%?

   A About 250 MW

   B About 500 MW

   C About 1,000 MW

   D About 5,000 MW

4. How much energy would be saved if the recycling rate of aluminum cans were doubled?

   F Approximately 14,000 MW

   G Less than 14,000 MW

   H Just over 15,000 MW

   J About 17,000 MW

5. Which recycled material currently produces the greatest energy savings?

   A Paper

   B Auto Batteries

   C Aluminum Cans

   D Tires

Passage III

Different types of electromagnetic radiation have different *wavelengths* (λ). All electromagnetic radiation travels at the speed of light (c = 3.0 10^8 m/s). The *frequency* of electromagnetic radiation is related to its wavelength by the equation $f = c/\lambda$. The units of frequency are the hertz (Hz). The magnitude is often expressed as gigahertz (1GHz = 10^9 Hz). The higher the frequency of the electromagnetic radiation, the more energy the wave has.

In the 1940s, a scientist discovered that a particular kind of electromagnetic radiation caused the rotation of water molecules. Research determined the reason—the polar nature of the water molecule. The water molecule has a partially positive charge on the hydrogen atoms and partially negative charge on the oxygen atoms. When placed in an alternating electromagnetic field, the molecules begin to rotate, trying to orient themselves to the field. This rotation heats the material containing the water.

Further research was conducted to determine the range of frequencies that would produce this effect. Table 1 and Figure 1 show the data obtained. Figure 2 shows the electromagnetic spectrum.

Table 1

Frequency (GHz)	Wavelength (m)
0.43	69.8×10^{-2}m
0.92	32.6×10^{-2}m
2.45	12.2×10^{-2}m
5.80	5.2×10^{-2}m

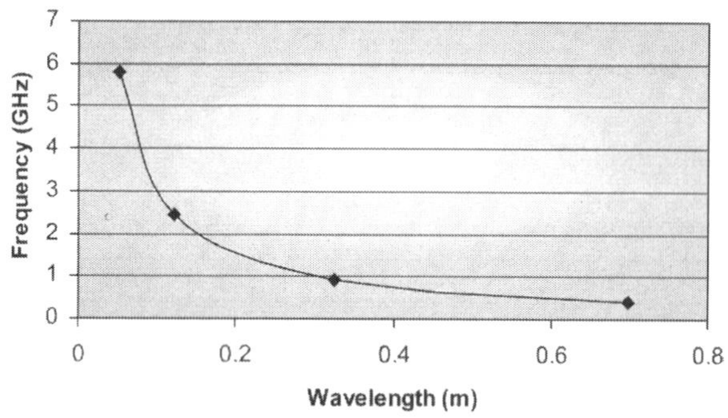

Figure 1

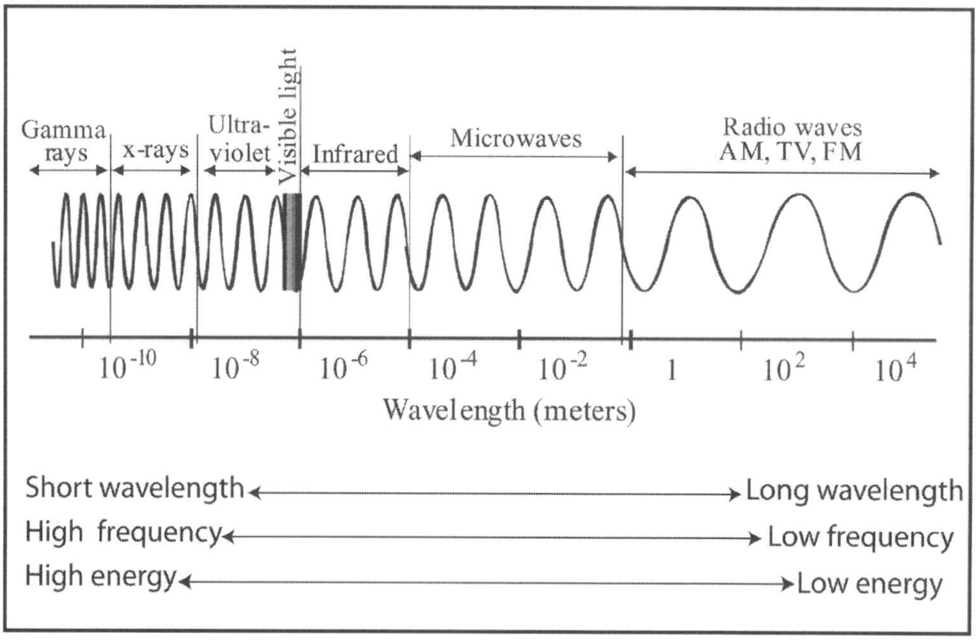

Figure 2

1. Use Table 1 and Figure 1 to describe the relationship between frequency and radiation.

   A Frequency increases with increasing wavelength.

   B Frequency decreases with increasing wavelength.

   C Frequency equals 3× wavelength.

   D Frequency equals 1/3 of wavelength.

2. Use Table 1 and Figure 2 to describe the type of electromagnetic radiation the scientist was studying.

   F Radio waves

   G Microwaves

   H Infrared waves

   J Ultraviolet radiation

3. Using Table 1, determine the highest energy radiation tested by the scientist.

   A 0.43 GHz

   B 0.92 GHz

   C 2.45 GHz

   D 5.80 GHz

4. One researcher used 3 GHz radiation. What was the wavelength of that radiation?

   F Greater than 70 cm

   G Between 12 and 32 cm

   H 10 cm

   J 1 cm

5. Research was conducted on electromagnetic radiation with wavelengths between 1 and 10 meters. These wavelengths of electromagnetic radiation did NOT rotate water molecules. Which of the following could be the reason for this result?

   A Radiation of this wavelength does not have enough energy to rotate water molecules.

   B Radiation of this wavelength is too energetic to rotate water molecules.

   C Radiation of this wavelength has too high a frequency to rotate water molecules.

   D Radiation of this wavelength is not traveling fast enough to rotate water molecules.

Chapter 3
Research Summaries

It is in the nature of humans to question. This process of inquiry is the central point of all scientific endeavors. The **Research Summaries** passages on the ACT test many of the same skills that Data Representation passages test—that is, your ability to work with graphs and tables and to detect trends in data.

RS passages have 6 questions

However, the RS passages also test your understanding of the process of scientific inquiry. More specifically, they will test your ability to understand the questions that others have asked, their research objectives, their experimental methods and their conclusions.

Know the Objective

A Research Summary passage contains a complete description of the research performed. Your first priority upon reading the passage is to know the purpose of the research. This is called the **objective**.

This is a statement that tells you what exactly the researchers were trying to find out. Sometimes, the objective is stated in terms of the researchers' hypothesis. A **hypothesis** is an educated guess about the outcome of an experiment.

Statement of Objective: Tells what the scientist intends to do.

Statement of Hypothesis: Tells what the scientist thinks will happen.

> ***Test Tip:*** Underline the objective.

Practice Questions 1

Practice Passage I

Students were studying the width of leaves in the forest. They collected and measured the leaves of redbud trees. They measured the width of the leaf from outer left edge to outer right edge at the widest point in centimeters. Their teacher wanted to confirm that sapling redbud trees have wide leaves while mature redbud trees have narrower leaves.

1. Which of the following statements illustrates the hypothesis of the study?

A The width of the leaf is measured between the outer edges.

B The collection and measurement of redbud leaves is time consuming.

C The students intended to compare redbud leaves on the same tree.

D Sapling redbud leaves are measurably wider than mature redbud leaves.

The best answer is D. This question is asking for a hypothesis or a predicted outcome of the experiment. Choice A describes a procedure or data collecting step done during the experiment. Choice B represents a critique of possible experimental flaws. Choice C is just plain incorrect. The students were not comparing leaves on the same tree. They were comparing the leaves of different aged trees. Therefore, Choice D is the best answer.

2. Which of the following statements represents the objective of the study?

A The width of the leaf is measured from the outer left edge to the outer right edge.

B The collection and measurement of redbud leaves is time consuming.

C The students intended to compare redbud leaves from trees of different ages.

D Sapling redbud leaves are measurably wider than mature redbud leaves.

3. Which statement below might represent a similar hypothesis?

A Distinct differences in the height of saplings can be identified and studied.

B Maple trees grown in a sunny environment will have higher sugar content in their leaves than maple trees grown in a shady environment.

C The number of oak trees has rapidly increased in the past few years representing a severe threat to biodiversity.

D Students were studying the ratio of saplings to trees in a mature hardwood forest.

UNDERSTAND THE PROCEDURE

The passage may give one or more procedures.

- If there is only one procedure, you must understand what it is.
- If there is more than one procedure, you must understand how each procedure is different from the others. What unique thing does it seek to test?

> ***Test Tip:*** Quickly jot down a brief description of the procedure in the margin. Use only key words. If there are two or more procedures discussed in the passage, use key words that describe the differences in the procedures. If the procedure is short, quickly do a mental summary.

Let's expand our redbud leaf study to include more procedures. While reading this procedure, be sure to underline the objective or hypothesis, and jot a quick description of the procedure in the margin.

Practice Questions 2

Practice Passage II

Students were studying the width of redbud leaves in the forest. They selected 10 plots in a forest and measured one meter cubes with string. Each of the plots received similar amounts of light, had similar foliage and had a similar terrain. They measured the widths of the tree trunks in the cube, to classify them as young (sapling) or older (mature) redbud trees. They collected leaves from four trees inside each cube (two sapling and two mature). Ladders were used to collect leaves from taller trees. Three leaves were randomly selected from each tree. The collected redbud leaves were measured. They measured the width of the leaf from outer left edge to outer right edge at the widest point in centimeters. Their teacher wanted to confirm that sapling redbud trees have wide leaves while mature redbud trees have narrower leaves.

1. Which statement below correctly summarizes the experimental procedure?

A Students collected leaves randomly throughout the forest and measured their width.

B Students collected and measured four leaves from each of 10 equal forest plots.

C Students measured leaf width from an equal number of sapling and mature redbud trees collected from similar forest plots.

D Students used ladders to climb the tallest redbud trees.

The best answer is C. Choices A, B and D are only one step in the procedure of this experiment. This question asks for a summary of all the procedural steps in this experiment. Therefore, Choice C is the best answer.

2. What key words might be used to describe this procedure?

   A Leaf width, half sapling/half mature, 10 plots/m^2

   B Tall redbud, 10 forests all mature

   C Confirm, sapling/mature, leaves

   D Outer/inner width, sapling/mature, 10/4/3

3. What is this experiment trying to test?

   A The difference in leaf width between young and old redbud trees.

   B The difference in width of young and old redbud tree trunks.

   C The difference in number of leaves in young and old redbud trees.

   D The difference in the number of young and old redbud trees in different forest plots.

Identify the Variables

A **scientific experiment** is designed to give **data** (measurable results). The data may be qualitative or quantitative. It must be meaningful, in terms of the objective. For example, recording height of nearby maple trees is not meaningful to a study of changes in redbud leaf width with tree age. Age of the redbud tree and width of the leaf are meaningful data.

To gather meaningful data, the experiment must be set up to examine only one condition (or **variable**) at a time.

There are three types of variables seen in an experiment:

- **independent variable** (sometimes called the **manipulated variable**) – The factors that are changed or manipulated during the experiment. They are the ones that the experiment is trying to test.
- **dependent variable** (sometimes called the **responding variable**) – This is usually the factor that the experimenter is measuring or counting. The dependent variable is the one that changes in response to the independent variable.
- **control variables** – All the other factors in the experiment. These are things that you attempt to control and are kept constant during the course of the experiment.

Scientists typically conduct experiments in batches. They do this to ensure that their data is **reproducible** (repeatable) and consistent. They group together the things that are to be tested the same way and run the experiment. The **experimental group** is the group that will be tested, such as 20 sapling redbuds and 20 mature redbuds in 10 similar forest plots. The **control group** is the group that the experimental group will be tested against.

Test Tip: Write out each variable in the margin. Pay particular attention to the control variable(s) and/or control group(s) if there are any.

Practice Questions 3

While rereading this experimental procedure, practice using the testing tips described in this chapter.

Practice Passage III

Students were studying the width of redbud leaves in the forest. They selected 10 plots in a forest and measured one meter cubes with string. Each of the plots received similar amounts of light, had similar foliage and had a similar terrain. They measured the widths of the tree trunks in the cube to classify them as young (sapling) or older (mature) redbud trees. They collected leaves from four trees inside each cube (two sapling and two mature). Ladders were used to collect leaves from taller trees. Three leaves were randomly selected from each tree. The collected redbud leaves were measured. They measured the width of the leaf from outer left edge to outer right edge at the widest point in centimeters. Their teacher wanted to confirm that sapling redbud trees have wide leaves while mature redbud trees have narrower leaves.

The data collected by the students is found in Table 1.

Table 1

Plot	Average Width of Sapling Leaves (cm)	Average Width of Mature Leaves (cm)
1	14.9	12.2
2	26.2	10.0
3	16.5	9.9
4	19.9	8.5
5	22.1	7.9
6	27.3	7.3
7	25.0	6.6
8	24.6	6.2
9	19.8	5.8
10	15.9	5.5

1. What is the independent variable in this experiment?

   A trunk width

   B leaf width

   C plot area

   D tree age

The best answer is D. Recall the different types of variables found in a scientific experiment. Independent variables are the variables that you manipulate or alter during the course of an experiment. In this example, the students changed the age of the redbud tree, young versus old. Choice A is the factor the students used to classify the age of each tree. Choice B is not the independent variable; it is the factor that is measured. Choice C is also incorrect. Plot area is one of the control variables mentioned in the text.

2. What is the dependent variable in this experiment?

   A trunk width

   B leaf width

   C plot area

   D tree type

3. Which of the following is NOT a control variable in this experiment?

   A tree type

   B plot area

   C trunk width

   D trunk height

4. What trend is apparent in the mature tree data?

   A Mature redbud trees have wider leaves than sapling redbud trees.

   B Sapling redbud trees have wider leaves than mature redbud trees.

   C The width of mature leaves is decreasing as plot number decreases.

   D Plot number increases as leaf width changes.

Research Summary Review

RS passages have 6 questions

- Test similar skills as Data Representation questions
- Must clearly understand scientific inquiry process
- Identify the experimental objective:
  - **objective** vs. **procedure**
- Summarize the procedure with key words
- Delineate the different variables:
  - **independent** - factor that is changed
  - **dependent** - factor that is measured
  - **controls** - factors that are kept constant

Chapter 3 Skill Review

Passage I

The *surface mass balance* (SMB) of an ice sheet is a way of measuring climate change. The SMB is the difference between accumulation and ablation. *Accumulation* is the addition of snow or ice to an ice sheet, while *ablation* is the removal of snow or ice. Scientists monitor accumulation and ablation on the ice sheet. The amount of each changes with elevation. When accumulation and ablation are equal, the SMB is zero. This is called the *equilibrium line* (EL).

Scientists test two different methods of determining the EL to determine which is more accurate.

Method 1 uses satellite data. Satellites designed to measure snow and ice are equipped with special sensors. Radiation is sent from the satellite to the surface, and the sensors record the radiation that is reflected from the surface back to the satellite. Frozen snow reflects more radiation than melting snow, so the radiation that gets reflected from the surface of the ice sheet will vary based on the conditions of the snow and ice. The data is input into a computer model for analysis. Computer models are designed to simulate current or future conditions based on available data.

Method 2 uses climate data which is then used as input into a computer model. Surface measurements of temperature and precipitation are collected from weather stations. Once inputted into the computer model, accumulation and ablation can be estimated. From these estimates, the SMB can be predicted.

Figure 1 compares the SMB from satellite and climate data along an elevation gradient of the ice sheet in west central Greenland for the year 2009. Elevation values are in meters above sea level (meters a.s.l.). Ice core data for the year 2009 was obtained a few years later. This data gave actual measurements of SMB values from the same area of the ice sheet. The ice core data is shown in Table 1.

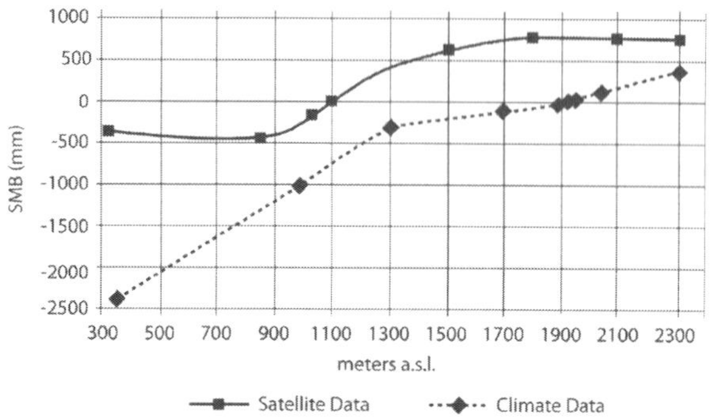

Table 1	
2009 Ice Core Data	
Elevation (meters a.s.l.)	SMB (mm)
300	-1950
500	-1800
700	-1520
900	-1100
1100	-610
1300	-450
1500	-195
1700	+370
1900	+510
2100	+590
2300	+700

1. What is the function of the ice core data?

   A It shows that the climate model data is entirely false.

   B It shows that the satellite data is entirely true.

   C It gives actual measurements with which the models can be compared.

   D It gives a third set of model values.

2. What effect does an increase in ablation have?

   F An increase in ablation decreases the SMB.

   G An increase in ablation increases the SMB.

   H An increase in ablation decreases accumulation.

   J An increase in ablation increases accumulation.

3. How does the measured value of the SMB differ between methods?

   A The SMB is greatest when estimated by climate data.

   B The SMB is greatest when estimated by satellite data.

   C The SMB is greatest when measured by ice cores.

   D The SMB is equal for all three methods.

4. A scientist hypothesizes that ongoing hail storms changed the reflectance patterns of the ice sheet surface. Which data would most likely be affected by this?

   F The satellite data

   G The climate data

   H The ice core data

   J All three data sets

5. According to satellite data, where is accumulation greater than ablation?

   A Below 1900 meters

   B Above 1000 meters

   C Above 1100 meters

   D Between 1100 and 2100 meters

6. What is the value of using the satellite or climate data models over actual ice core measurements?

   F The models can be more accurate than real measurements.

   G The models are less expensive than real measurements.

   H The models can be used to predict current and future trends.

   J The models require less interpretation than real measurements.

Passage II

A cloud is a mass of condensed water vapor located in the atmosphere. There are many types of clouds. They are classified based on their altitude and whether or not they are accompanied by precipitation. The altitude at which they form is not always the same. Each cloud type generally forms within a range of altitudes. Cloud formation depends on many factors, including the climate of a given area, the air pressure and water content of the air at the altitude that the cloud is forming.

Atmospheric scientists performed experiments to determine the predominant cloud formations over a given Gulf Coast land area in the month of May. They directed a laser upward at a 90° angle (θ_i) toward a cloud during formation. Upon striking the water vapor of the cloud, the laser light is affected in a variety of ways. Some is refracted, some is scattered and some is reflected. The reflected light rebounds to strike the surface of a light detector located at the laser delivery site. Based on the time it takes the reflected light to return, the altitude of the cloud is determined.

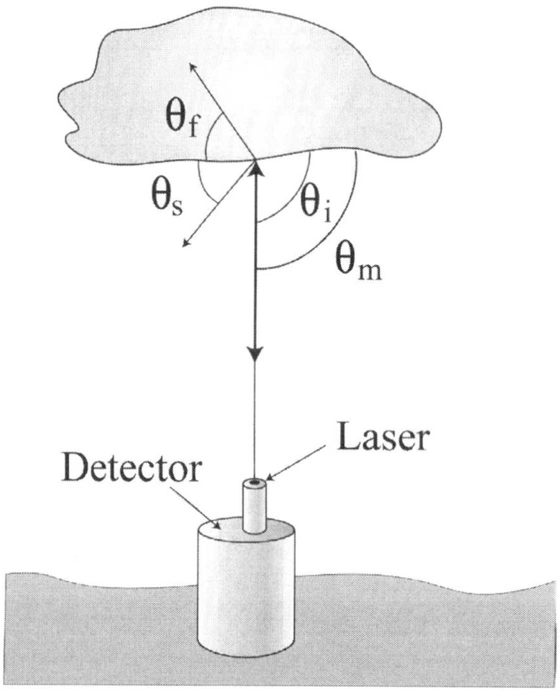

They then correlated the average altitude of each cloud to the incidence of rain as shown in Table 1.

Table 1

Cloud Type	Average Altitude (m)	Precipitation Cloud
Stratus	1800	yes, light
Cumulonimbus	2500	yes, heavy
Altostratus	4800	yes, light
Altocumulus	6000	no
Cirrostratus	6100	no
Cirrus	7200	no

1. Using Figure 1, what angle of light did the scientists use to measure altitude of the cloud base?

   A θ_f
   B θ_i
   C θ_s
   D θ_m

2. Which of the following graphs depicts the altitude data most clearly?

   F

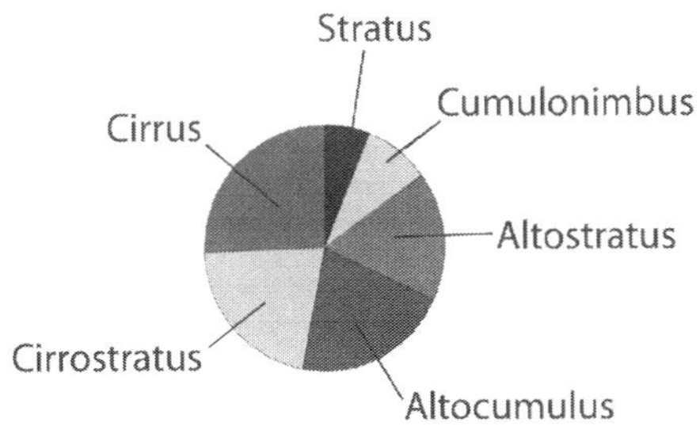

   G

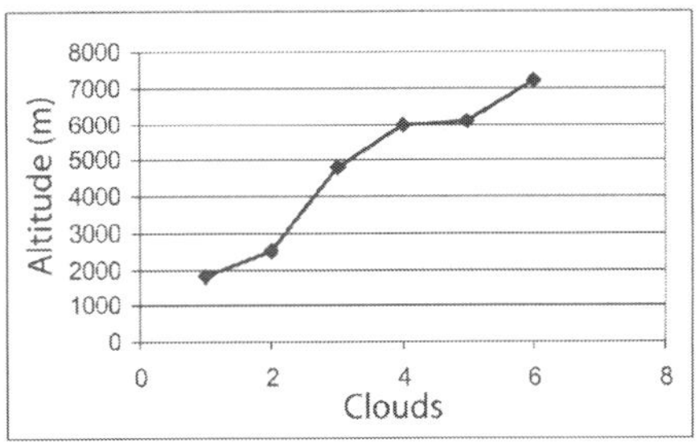

   H

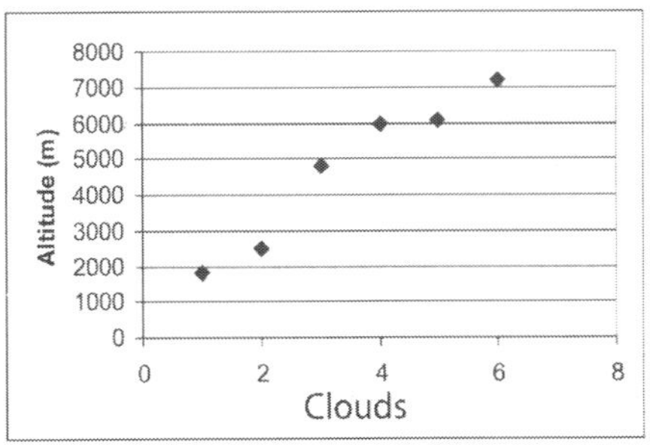

   J

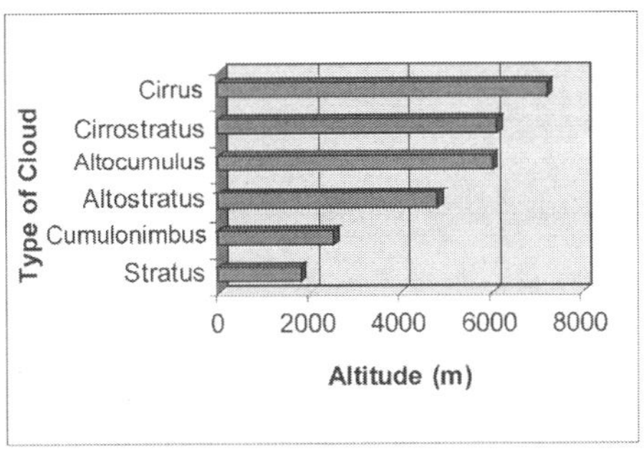

3. Clouds forming at altitudes above 6000 m are called high altitude clouds. The atmosphere is very cold at that altitude, so these clouds are made of varying degrees of ice crystals. Based on Table 1, which of the following clouds will likely have the highest percentage of ice?

   A Stratus
   B Altocumulus
   C Cirrostratus
   D Cirrus

4. One Gulf area received a record 10 inches (25 cm) of rain and thunderstorms in 2 days during the study period. What was the predominant cloud type present on those days?

   F Stratus
   G Cumulonimbus
   H Altostratus
   J Altocumulus

5. Based on Table 1, what would be a correct conclusion about clouds in the Gulf Coast study area?

   A No clouds form beneath 1800 meters.
   B No stratus clouds can be found at 2000 meters.
   C All cirrus clouds are found at 7200 meters.
   D Both altocumulus and cirrostratus clouds can be found at 6,050 meters.

6. Late in the study period, scientists decided to begin studying the density of clouds. Denser clouds should reflect more light than less dense clouds. Which of the following would describe a high-density cloud?

   F $\theta_s + \theta_m$ is much greater than θ_f.
   G $\theta_f + \theta_s$ is much greater than θ_m.
   H $\theta_s + \theta_m$ is much less than θ_f.
   J θ_i equals θ_m.

Passage III

A student conducted experiments on four different salts, using the following equipment:

four 250 mL beakers

deionized water

various salts (see Table 1)

hotplate stirrer

magnetic stirring bar

Table 1

Salt	Sample Mass (g)
$NaNO_3$	30 g
KNO_3	30 g
KCl	30 g
$Ce_2(SO_4)_3$	30 g

Experiment 1

He added the salt sample to a beaker containing 100 mL of the deionized water at 25°C. He placed the magnetic stirring rod in the beaker. He placed the beaker on the hotplate stirrer and inserted a thermometer into the salt solution. He turned on the magnetic stirrer and the heat. He recorded the temperature at which the salt dissolved completely.

Experiment 2

Next, the student added each salt to 100 mL of deionized water at 10°C. He added the salt until the solution was saturated. He then heated the solution by 10°C and added more salt, again until the solution was saturated. He recorded the mass of each of the four salts added at each temperature. His data are shown in Figure 1.

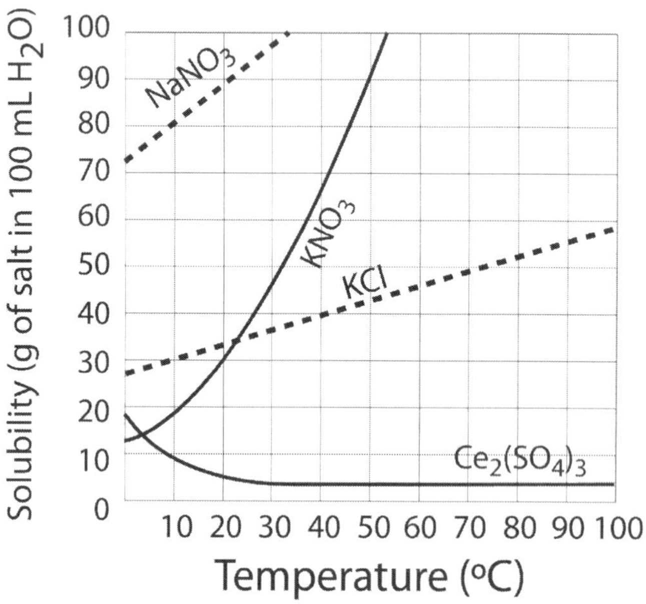

1. What was the objective of Experiment 1?

   A To determine the solubility of each of the four salts at 25°C

   B To determine the temperature required to dissolve 30 grams of each salt in 100 mL of water

   C To determine the amount of water required to dissolve 30 grams of each salt in 100 mL of water

   D To determine how many different salts will dissolve in 100 mL of water

2. Which of the following correctly shows the student's experimental set-up?

F

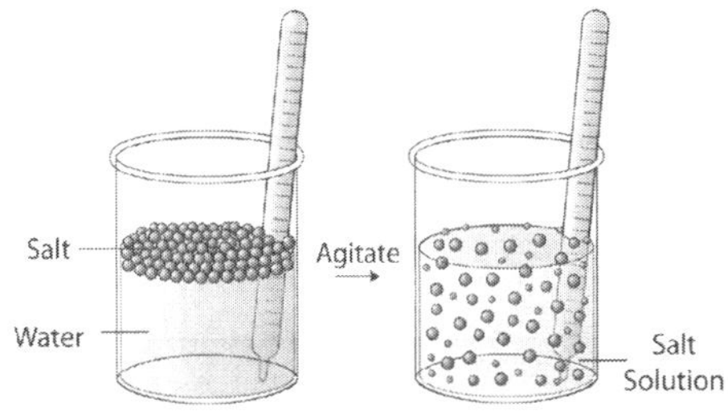

G

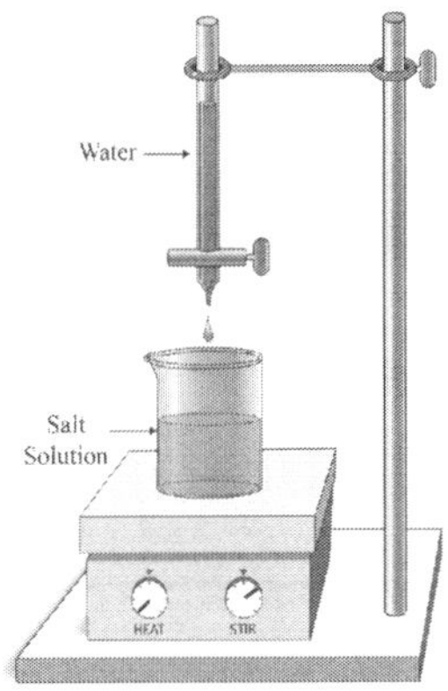

H

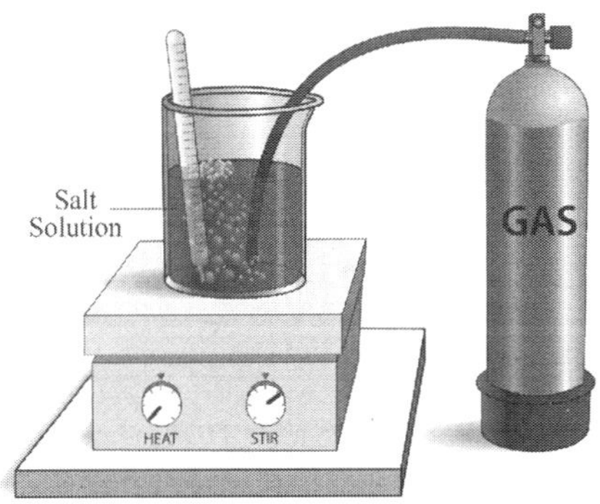

J

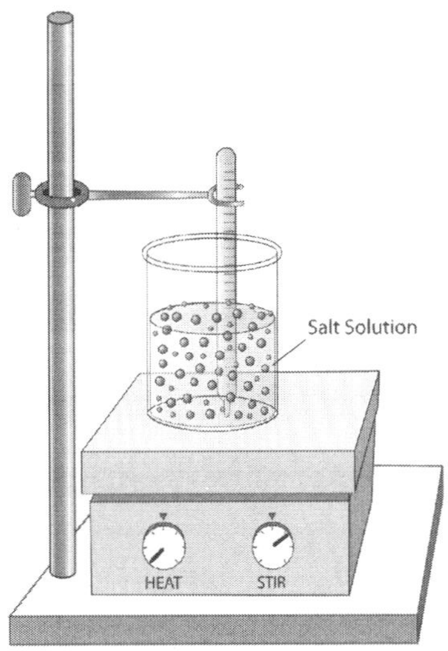

3. Using Figure 1, what was the only salt solution that was saturated at 30 g/100 mL and 25 °C?

   A $NaNO_3$

   B KNO_3

   C KCl

   D $Ce_2(SO_4)_3$

4. During the course of Experiment 2, the student discovered that 100 grams of one salt dissolved in 100 mL of water at a temperature under 50°C. Which salt was it?

   F $NaNO_3$

   G KNO_3

   H KCl

   J $Ce_2(SO_4)_3$

5. The apparatus the student used supports which of the following hypotheses?

   A Magnetism dissolves salt.

   B Heat and stirring help salts dissolve.

   C Salts turn into liquids as the temperature increases.

   D Stirring and heat will keep salts in the solid state.

Use the figure below to answer question 6.

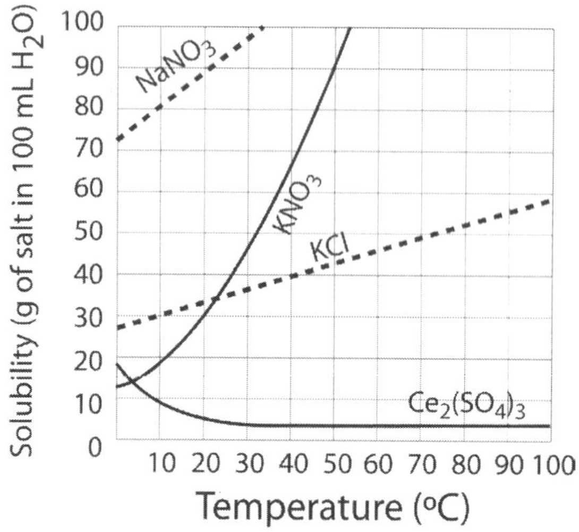

6. According to the graph, how is $Ce_2(SO_4)_3$ different from the other compounds shown?

   F The solubility of $Ce_2(SO_4)_3$ does not change with temperature.

   G The solubility of $Ce_2(SO_4)_3$ decreases when the temperature is increased.

   H All of the other compounds are more soluble at 0°C.

   J All of the other compounds are less soluble at 10°C.

Chapter 4
Conflicting Viewpoints

A **viewpoint** is a position or perspective from which an issue is considered. Your viewpoint is influenced by your own knowledge, experiences and opinions. It is influenced by your family and your friends. It is influenced by where you grew up and where you live now. It is influenced by the media that you see and listen to. No two people on Earth have exactly the same viewpoint on every issue.

CV passages have 7 questions

The same is true of scientists. They bring their different perspectives to their work. This may mean that they ask different questions, research different aspects of a problem, conduct different kinds of experiments or analyze data differently. Based on these differences in their methods or analysis, scientists often reach different conclusions. They have **Conflicting Viewpoints**.

Analyzing a Viewpoint Part 1

The ability to analyze a person's viewpoint and to see how it deviates from the viewpoints of others is tested by Conflicting Viewpoints (CV) passages. These passages describe the viewpoints of 2 or more scientists, researchers or theories concerning a scientific topic. There are two things that you should do immediately when you recognize a CV passage:

- **Underline the sentence that summarizes each viewpoint.** This is usually, but not always, the first sentence. At least one CV question will ask you to summarize the central point of one or both scientists.
- **Number the points that each scientist makes to support his viewpoint.** Place a number next to each point that each scientist makes. Make corresponding points have corresponding numbers. This numbering system will allow you to see if there are any points that only one scientist addressed. If so, there will surely be a question about that point. It will also allow you to quickly locate a particular topic within each scientist's segment.

Often times, the source of disagreement will have already been solved. A great example is the Geocentric vs. Heliocentric debate.

Practice Questions 1

Practice Passage I

Scientists and philosophers between the 1500s and the 1700s debated the pattern of movement of the Earth, Sun, Moon, stars and other planets. Two primary theories are explained below.

Theory I

The Geocentric Theory was a core scientific belief from about 5 B.C. until the late 1600s. This theory stated that the universe revolved around Earth. Two common observations were believed to support this idea.

The first is that the Sun, stars and other planets actually appear to revolve around the Earth each day. Using the naked eye, one can see the stars track their passage through the sky each night. The stars around the poles travel in a fairly tight circle. Stars nearer the equator rise and set each day, circling back to their rising point.

The second is the common sense perception that the Earth is solid and stable. Since humans never feel the Earth move, it seemed obvious to many that it was not, in fact, moving. If the Earth is at rest and other objects in space are moving in a predictable pattern, it follows that they must be moving around the Earth.

Theory II

Proponents of this theory stated that the universe actually revolved around the Sun, rather than the Earth. This was called the Heliocentric Theory. The evidence to support this theory was both observational and geometric. Although the Sun, stars and planets do rise and set, the pattern in which they do so changes from day to day, month to month and year to year. No simple pattern of revolution could account for the gradual slide of the constellations through the sky.

Nicholas Copernicus spent many years in careful astronomical observation, followed by the geometric calculations. The result was a text of over 400 pages that contained tables and equations that would predict the position of a star at a date in the future or a date in the past.

1. **Which of the following pieces of evidence seems to support the Geocentric Theory?**

   A Humans feel the Earth move during an earthquake.

   B Humans do not feel the Earth moving through space.

   C The stars have a different pattern of movement than the Sun.

   D The days are longer in the summertime and shorter in winter.

The best answer is B. Careful examination of the passage reveals that Theory 1 is the Geocentric Theory; it uses two common observations as support: (1) the Sun and other stars appear to revolve around the Earth, and (2) the Earth is solid and stable. Choice A is incorrect because earthquakes are never mentioned in the text. Both Choices C and D support the Heliocentric Theory.

2. **A heliocentrist would most likely argue that:**

   A the Earth revolves around the Moon.

   B the Moon revolves around the Sun.

   C the Earth revolves around the Sun.

   D the Sun revolves around the Earth.

3. **Both theories contend that:**

   A the Earth moves.

   B the Sun moves.

   C stars move.

   D everything stays still.

4. **The planets in other solar systems revolve around other central stars. Which of the theories accounts for this fact?**

   A Only the Geocentric Theory

   B Only the Heliocentric Theory

   C Both theories

   D Neither theory

After seeing these common CV questions, let's talk about a couple things you can do to better answer them.

- **Look for key words and phrases to help you narrow down your answer choices.** For the Heliocentric Theory, the key phrase is "revolves around the Sun." This phrase will clearly help you quickly answer question 2.
- **Be careful to assign a point to the correct theory.** Go back to your numbered points, and be sure that you have not switched them.
- **Look for answer choices that are clearly not true.** Use common sense. For instance, "everything stays still" is a highly unlikely answer to any question about the stars.

Did you notice that last question was sneaky? Well, sometimes a Conflicting Viewpoint question will present you with a fact, and then ask which scientist's, researcher's or theory's conclusions would support that fact. To answer this one, go back to the points of each theory that you numbered. You should notice that neither theory said anything about another solar system. (This is another example of using key words.)

- **Pay attention when a question is asked about something that is not mentioned in the passage.** Since the ACT Science Reasoning Test does not assess your science knowledge, it does not generally expect you to be able to make assumptions about topics outside the passage.

Analyzing a Viewpoint Part 2

Sometimes a CV passage describes a scientific debate that has not yet been "solved." Remember that CV passages and questions are not designed to evaluate your ability to determine the correct or truthful side of the debate. They are designed to evaluate your ability to understand and analyze the debate itself. So, you should treat these questions the same way that you would a passage about a now-uncontroversial topic.

Practice Questions 2

Practice Passage II

Policy makers, scientists and environmentalists debate the use of nuclear power. Within each group, there are both proponents and opponents of nuclear power. Two scientists examine the topic.

Scientist 1

Nuclear power is a safe and environmentally responsible energy technology. Only two accidents have ever occurred in a commercial power reactor. And, only one of these accidents resulted in a breech of the containment system and release of radiation. In any case, both of these reactors were early designs, called Generation II reactors. The newer Generation III reactors being built today feature higher electricity output and many more safety features. In addition, Generation III reactors use much less water and produce far less nuclear waste than earlier reactor technologies, so they have far less impact on the environment. Finally, nuclear reactors produce no air pollution at all and do not contribute greenhouse gases to the atmosphere.

Scientist 2

The physical risk of using nuclear power far outweighs any perceived environmental benefits. Although there have only been two serious accidents at a nuclear plant, there have been many smaller accidents, occurring in both Generation II and Generation III reactors. It is simply luck that none of these accidents was larger. The fact is that no safety measures are perfect, and a nuclear reactor failure can have dire consequences. Besides the destruction resulting from the initial failure, the possibility of future health problems resulting from widespread radiation exposure must be considered. This level of risk is simply not present with any other power producing option.

Generation II reactors use about 100 million gallons of water a day. Generation III reactors use about 30 million gallons of water a day. So, although Generation III reactors use much less water, they still use significant amounts. And, although nuclear plants do not produce any air pollution like carbon dioxide, they do release large amounts of water vapor, which is in fact a greenhouse gas.

1. **What is the primary problem that Scientist 2 has with nuclear power?**

   A He feels it is unsafe.

   B He feels that it is ecologically damaging.

   C He feels that it makes electricity too costly.

   D He feels that it produces too much waste.

The best answer is A. The first sentence in Scientist 2's passage clearly defines his position on nuclear power. Choices B–D are not even mentioned in the text. Although these are legitimate concerns, they are not brought up in this passage. When answering these types of questions, carefully examine your numbered points within the passages.

2. **Based on Scientist 2's explanation, we can conclude that:**

   A nuclear reactors do release a greenhouse gas.

   B nuclear reactors release carbon dioxide.

   C nuclear reactors do not release water vapor.

   D nuclear reactors pollute the water supply.

3. **A Generation II Reactor typically produces 500 tons of radioactive waste per year. A Generation III Reactor produces about 40 tons. Which scientist could use this statement to support his position?**

   A Scientist 1

   B Scientist 2

   C Neither scientist

   D Both scientists

4. **Scientist 1 states that the water usage of a Generation III reactor is:**

   A equal to that of a Generation II reactor.

   B 1/3 of that of a Generation II reactor.

   C 3x that of a Generation II reactor.

   D far less than that of a Generation II reactor.

Questions 1–3 should have been fairly easy, if you have underlined the points of each scientist and analyzed them with the strategies we have discussed so far. Question 4 brings us to the final point:

- **As with all ACT passages, pay close attention to the numbers.** There will often be a question that asks you to interpret statements about raw numbers, fractions or percentages. This is the case in question 4 above.

Conflicting Viewpoint Review

- Summarize the **main idea** within each viewpoint.
- Pick out particular points or facts used to support each viewpoint:
  - – look for **key words**
  - – carefully reexamine all the points
  - – watch out for **false statements**
  - – use caution when analyzing raw data

Chapter 4 Skill Review

Passage I

Scientists and philosophers throughout history have questioned the nature of light. Classical physics deals with two categories of phenomena: particles and waves. Particles are objects that have mass and take up space. Waves, such as waves in the ocean or sound waves, are disturbances that repeat regularly in space. Waves require a medium to travel from one place to the next but do not result in the net movement of matter.

Light exhibits the properties of reflection, diffraction and refraction. Reflection is the bouncing back of light once it hits a boundary. Diffraction is the bending of light as it passes around an obstacle or though a narrow slit. Refraction is the bending of light as it passes from one transparent medium to another, for example, from air to water. Following are three alternative explanations of the nature of light.

The Particle Theory of Light

In 1704, Isaac Newton proposed that light is composed of streams of particles. These particles are emitted in all directions from a source and travel in straight lines. This idea was soon replaced by the wave theory of light.

The Wave Theory of Light

In 1690, Christian Hoydens published his proposal that light is composed of waves that are emitted in all directions from a source. He proposed that the waves travel through a medium called the luminiferous ether. This idea gained support through several experiments following its development.

The Quantum Mechanical Theory of Light

This theory was proposed by Albert Einstein in 1905 and is the theory most widely supported today. This theory holds that light has a "dual nature," acting simultaneously as both a particle and a wave. According to this theory, light consists of beams of massless particles called photons that can travel through empty space. These particles carry energy. Photons exhibit all the properties of waves, as well as all the properties of particles. For example, photons can be refracted, like waves. They behave as particles in that they can actually eject electrons from metal surfaces, in the same manner that a rapidly moving object will displace other objects in its path. Photons' ejection of electrons from atoms is known as the photoelectric effect.

1. **Reflection is a phenomenon observed in both particles and waves. Which of the theories accounts for the observation that light is reflected?**

   A The particle theory only

   B The quantum mechanical theory only

   C Both the particle and the wave theory, but not the quantum mechanical theory

   D All three theories account for this observation

2. **Refraction has only been observed in waves. Which of the theories accounts for the refraction of light?**

   F The wave theory only

   G The quantum mechanical theory only

   H Both the wave theory and the quantum mechanical theory

   J All three theories account for this observation.

3. In the late 1800s, an experiment discredited the presence of a luminiferous ether. Which theory was likely most weakened by this development?

A The particle theory

B The wave theory

C Both the particle and the wave theory

D The quantum mechanical theory

4. Solar panels work because sunlight ejects electrons from certain atoms contained in the panels. The electrons are then routed through a circuit to make electricity. According to the information in the previous passages, which of the theories above is LEAST consistent with the present-day operation of solar panels?

F The particle theory

G The wave theory

H The quantum mechanical theory

J All theories are inconsistent with this phenomenon.

5. The quantum mechanical theory of light is most consistent with which of the following ideas?

A Waves and particles are separate and unrelated phenomena.

B Light requires a medium to be transmitted.

C Mass can neither be created nor destroyed.

D Matter can exhibit wave-like properties.

6. Classical physics held that particles are composed of matter, and all matter has mass. Which of the theories above is LEAST consistent with this idea of classical physics?

F The particle theory

G The wave theory

H The quantum mechanical theory

J All theories are inconsistent with this view.

7. One of Newton's arguments against the wave theory of light was that light did NOT exhibit diffraction. What experiment likely weakened Newton's argument the most?

A Thomson's experiment with the photoelectric effect

B The Michelson-Morley experiments which cast doubt on the presence of the luminiferous ether

C Thomas Young's experiment in which he showed that light bent while traveling through a small slit

D Faraday's experiment documenting the effects of a magnetic field on light

Passage II

Scientists debate the origins of Earth's Moon. The Moon's density is very different than that of Earth, suggesting the two have very different compositions. The Earth is composed of a very dense liquid iron core surrounded by a somewhat less dense semi-molten rock mantle and a low-density solid crust. The low density of the Moon suggests that it does not have a significant iron core and that it is composed of less dense materials throughout. Two alternative explanations for the formation of the Moon are presented below.

Explanation 1: "Capture Theory"

The Moon was formed somewhere in the solar system, far away from Earth, where iron was less abundant. The Moon was then captured by Earth's gravitational field as it traveled past the Earth sometime after its formation.

Explanation 2: "Giant Impactor Theory"

A planetesimal (small planet) the size of Mars struck the Earth early in its history. The impact generated enormous amounts of heat and sent huge quantities of molten debris from Earth's crust and mantle into Earth's gravitational field. This debris formed an orbiting ring around Earth, and the material in this ring eventually fused to form the Moon. The Moon was likely almost entirely molten at the time of its formation. Collisions of this nature are believed to have been fairly common in the early solar system.

Supporters of the Giant Impactor Theory believe it very unlikely that Earth could have captured the Moon into its present orbit. The Moon would have been traveling past Earth at a very high velocity, thus making it virtually impossible that Earth could have captured it.

1. **Which explanation(s) accounts for the differing densities of the Earth and the Moon?**

   A Capture Theory only

   B Giant Impactor Theory only

   C Both Capture Theory and Giant Impactor Theory

   D Neither Capture Theory nor Giant Impactor Theory

2. **Identical relative abundance of oxygen isotopes imply that the Earth and Moon were formed the same distance from the Sun in the solar system. Which theory is best supported by this data?**

   F Capture Theory only

   G Giant Impactor Theory only

   H Both Capture Theory and Giant Impactor Theory

   J Neither Capture Theory nor Giant Impactor Theory

3. **The low presence of volatile compounds such as water in lunar rocks suggests that lunar rocks have been exposed to excessive heating compared to rocks of Earth. How would supporters of the Giant Impactor Theory likely explain this data?**

   A The Earth and Moon's close proximity to the sun resulted in evaporation.

   B The capture of the Moon generated a large amount of heat.

   C The heat from the Moon's core caused evaporation.

   D The heat from the impact melted the rock that became the Moon.

4. **Supporters of the Giant Impactor Theory believe the planetesimal's collision with Earth was off-center, rather than straight-on, thus causing it to miss Earth's core. The angular momentum of the Earth-Moon system supports this suggestion. This suggestion is also supported by:**

   F the lower density of the Moon compared with that of the Earth.

   G the identical ages of the Earth's and Moon's rocks.

   H the low water content of the Moon's rocks.

   J the frequency of collisions in the early solar system.

5. **Both explanations are consistent with which of the following ideas?**

   A The Earth and Moon were formed at approximately the same place in the solar system.

   B Collisions and near-collisions of objects in the early solar system were not uncommon events.

   C The Earth and Moon appear to have been formed in very similar manners.

   D Objects in the early solar system moved with a high enough velocity to make capture in another's gravity unlikely.

6. **In recent years, the Capture Theory, once the leading explanation of Moon formation, has lost popularity in favor of the Giant Impactor Theory. Which of the following events likely resulted in this shift?**

   F Calculations of the Earth's and Moon's densities

   G Discovery of the Earth's core

   H New evidence indicating the frequency of collisions and near-collisions in the early solar system

   J New calculations revealing the gravitational force necessary for the Earth to capture the Moon

7. **Which piece of evidence do supporters of the Giant Impactor Theory use to refute the Capture Theory?**

   A The density of the Moon vs. density of the Earth

   B The velocity of a passing Moon

   C The frequency of collisions during the formation of the solar system

   D The abundance of iron in Moon rocks

Practice Test

SCIENCE TEST

35 Minutes—40 Questions

DIRECTIONS: There are seven passages in this test. Each passage is followed by several questions. After reading a passage, choose the best answer to each question. You may refer to the passages as often as necessary.

You are ***NOT*** permitted to use a calculator on this test.

Passage I

Scientists working in a protected wildlife reserve collect population data for various bird species each year. The population of six bird species was counted in 2005 and again in 2010. The population data for these species are displayed in pie graphs, one for each year (Figures 1 and 2). Because of their diet, these birds require the soil in their habitat to leave a great deal of organic matter. Soil organic matter is dependent on the availability of water in the soil. The scientists do not measure organic matter directly. However, they can estimate the water content in soil based on the average rainfall. Three weather stations—X, Y and Z—are set up within the wildlife reserve, 20 kilometers apart from each other. Rainfall data was collected from each weather station over a six-year period, from 2004 to 2010 (Table 1). Figure 3 shows the relationship between soil water content versus rainfall. The required water content of the soil is indicated for each bird species in Table 2.

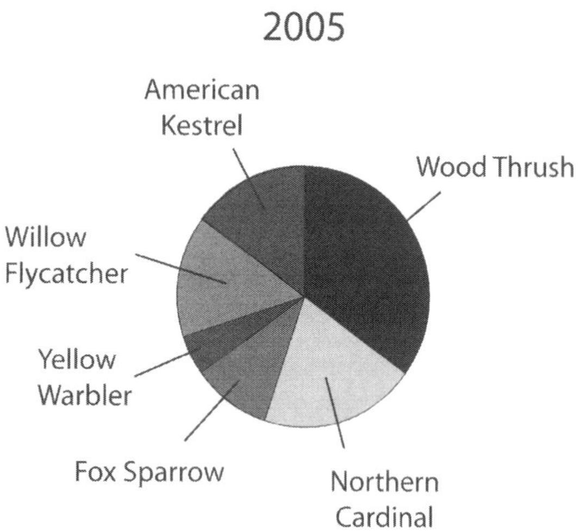

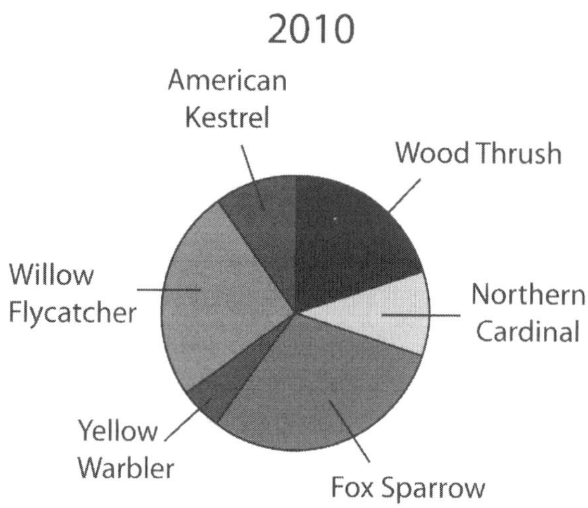

Table 1

Rainfall (millimeters)			
Year	Station X	Station Y	Station Z
2004	760	824	785
2005	790	750	699
2006	642	725	702
2007	590	645	610
2008	209	301	225
2009	199	185	155
2010	298	345	206

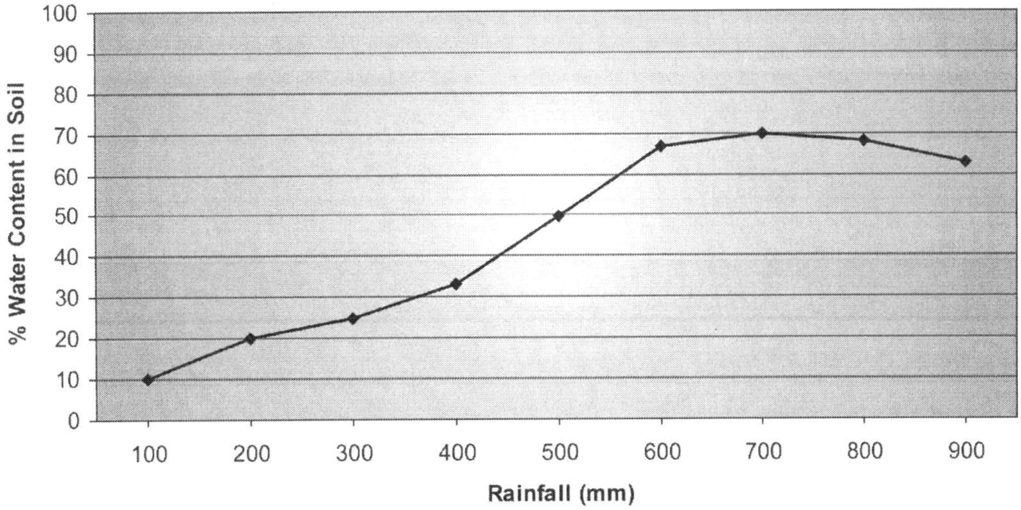

Table 2	
Minimum Water Content of Soil (%) Required for Survival	
Bird Species	% Water Content
Wood Thrush	70
Northern Cardinal	65
Fox Sparrow	20
Yellow Warbler	20
Willow Flycatcher	30
American Kestrel	50

1. **In 2005, about what percent of the population was the Northern Cardinal?**
   - **A** 5%
   - **B** 20%
   - **C** 50%
   - **D** 75%

2. **Between 2005 and 2010, how did the population of the Fox Sparrow change?**
   - **F** It decreased by about 5%.
   - **G** It decreased by about 15%.
   - **H** It increased by about 5%.
   - **J** It increased by about 15%.

3. **Which bird species was most negatively impacted by the change in rainfall over this period?**
   - **A** Yellow Warbler
   - **B** Willow Flycatcher
   - **C** Wood Thrush
   - **D** Northern Cardinal

4. **According to Figure 3 and Tables 1 and 2, in what year was rainfall below the minimum required by the Wood Thrush?**
   - **F** 2007
   - **G** 2008
   - **H** 2009
   - **J** 2010

5. **Which of the following statements best explains how rainfall affected the Fox Sparrow over this period?**
   - **A** Its population increased because only one year received inadequate rainfall.
   - **B** Its population increased because there were no years of inadequate rainfall.
   - **C** Its population decreased because several years received too much rainfall.
   - **D** Its population decreased because all the years received inadequate rainfall.

Passage II

Soil moisture is used to assess water needs for irrigation, to control flooding and to manage overall water supply. Using precipitation and evapotranspiration rates, an annual budget of soil moisture can be established. This is called the *water balance*. The water balance of a soil is determined by calculating the inputs, outputs and storage changes of water at the surface. The major input of water is from precipitation (P), while the major output is evapotranspiration. Actual evapotranspiration (AE) is used water, or the amount of water that is actually evaporating and transpiring. It increases as temperature increases, as long as there is water available. Potential evapotranspiration (PE) is the demand for water, or the amount of water that would be evaporated under optimal conditions with an unlimited supply of water. The soil moisture storage (ST) is the amount of water held in the soil. Adding or using water from storage changes the amount stored (DST). The maximum amount of water the soil can hold is the field capacity. Surplus (S) occurs when the supply meets or exceeds the demand for water and the soil is saturated. Deficits (D) occur only when the soil is completely dry (the soil has utilized all the stored water) and the demand for water exceeds the supply. When supply meets or exceeds the demand once again, the soil is no longer in deficit and begins to recharge.

Table 1 shows annual water balance data for a soil. All measurements are in millimeters (mm). The field capacity of the soil is 90 mm. A diagram of the climatic data and soil water balance for the same location is shown in Figure 1.

Table 1

Soil Field Capacity = 90 mm

	Jan	Feb	Mar	Apr	May	Jun	Jul	Aug	Sep	Oct	Nov	Dec
P	50	48	60	78	100	106	82	84	86	73	56	45
PE	0	0	5	40	84	123	145	126	85	44	8	0
P-PE	50	48	55	38	26	-17	-63	-42	1	29	48	45
DST	0	0	0	0	0	-17	-63	-10	1	29	48	12
ST	90	90	90	90	90	73	10	0	1	30	78	90
AE	0	0	5	40	84	123	145	94	85	44	8	0
D	0	0	0	0	0	0	0	32	0	0	0	0
S	50	48	55	38	26	0	0	0	0	0	0	33

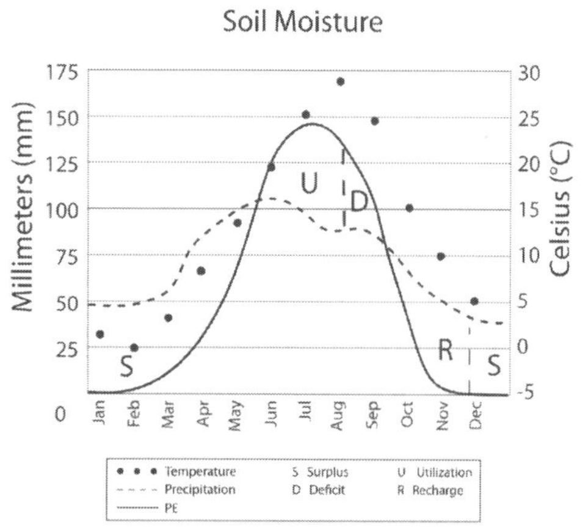

6. **In November, precipitation equals 56 mm, potential evapotranspiration equals 8 mm and the water storage is 78 mm. Which of the following statements is true?**

   F The soil is in surplus.

   G The soil is in utilization.

   H The soil is in recharge.

   J The soil is at field capacity.

7. **How are temperature and potential evapotranspiration (PE) related?**

   A High temperatures increase the demand for water.

   B High temperatures decrease the demand for water.

   C Low temperatures increase the demand for water.

   D Temperature does not affect the demand for water.

8. **In what month is the soil driest?**

   F June

   G July

   H August

   J September

9. **In December, 45 mm of precipitation fell. What happened to this precipitation?**

   A 12 mm was taken from storage and 33 mm was surplus.

   B 12 mm was added to storage, reaching field capacity, and 33 mm was surplus.

   C All 45 mm was added to storage, reaching field capacity.

   D All 45 mm was taken from storage to meet PE demands.

10. **If PE is greater than AE and precipitation, and the soil is dry, which of the following is true?**

    F Soil water recharge is likely to occur.

    G Soil water surplus is likely to occur.

    H Soil water utilization is likely to occur.

    J Soil water deficit is likely to occur.

Passage III

A group of researchers investigated the ecological consequences of predator behavioral response to global climatic variation. Their study evaluated the dynamics of a three trophic level system in Wisconsin, where wolves limit abundance of moose. Moose limit productivity of fir trees. Wolves have been found to hunt in larger packs during years of increased winter snows associated with the *North Atlantic Oscillation* (NAO). The NAO moves between positive and negative phases. A negative NAO was associated with increased snow accumulation; whereas, a positive NAO was associated with decreased snow accumulation. Figure 1 shows wolf and moose abundance in Wisconsin. Figures 2 and 3 show how climate influenced the ecosystem from wolf behavior to growth of fir trees. Density was measured as the number of animals over the area studied. Pack size was measured as the number of wolves per pack.

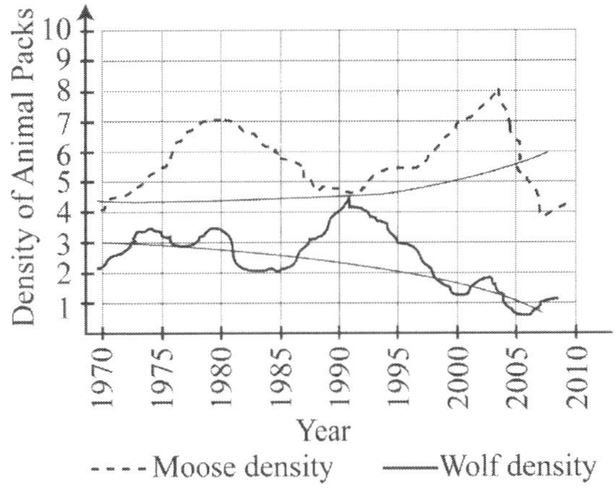

Figure 1

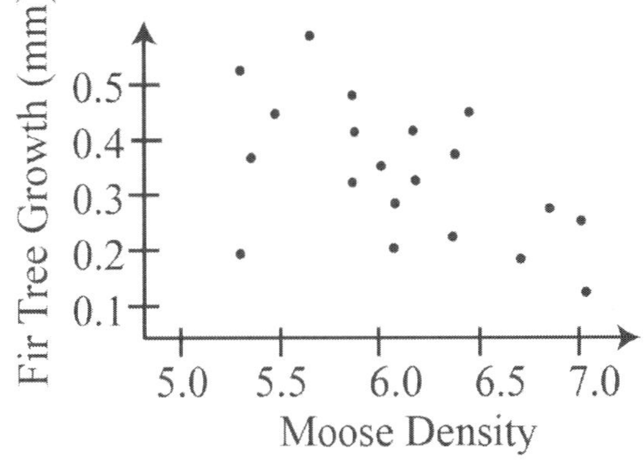

Figure 2

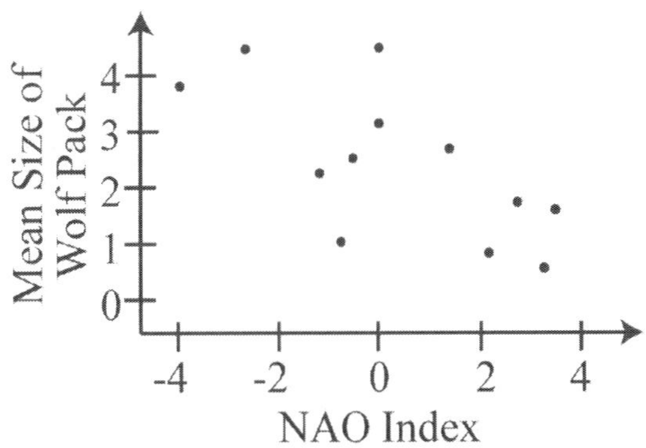

Figure 3

11. What is the relationship between NAO index and wolf pack size during the study period?

A As the NAO index becomes positive, wolf pack size increases.

B As the NAO index becomes negative, wolf pack size increases.

C As the NAO index becomes negative, wolf pack size decreases.

D NAO is not well correlated with wolf pack size.

12. Which of the following best describes the relationship between wolf density and moose density?

F Moose winter mortality was dependent on wolf pack size.

G Moose winter mortality was unrelated to wolf pack size.

H Wolf pack size was dependent on moose density.

J Wolf and moose density are positively correlated.

13. How do moose affect the productivity and growth of fir trees?

A Moose increase productivity and growth of fir trees.

B Moose inhibit productivity and growth of fir trees.

C Moose increase productivity but inhibit growth of fir trees.

D Moose do not affect overall productivity and growth of fir trees.

14. What was the overall trend in moose density during the study period?

F Negative

G Positive

H A flat trend

J No clear trend

15. Based on the passage and figures, which of the following statements is true?

A The density of wolves is dependent on the density of fir trees.

B The NAO index is dependent on snow accumulation.

C Fir tree growth is directly dependent on wolf pack size.

D The density of moose is dependent on the density of wolves.

Passage IV

Two students studied earthworms using the following equipment:

- shovel
- fine wire screen sieve
- wheelbarrow
- five 0.5 m^2 plots in forest
- five 0.25 m^2 plots in field
- meter ruler
- triple beam balance

Procedure:

Students selected five plots in the forest and five plots in the field that had a similar slope and received a similar amount of sunlight. Students then dug a 0.125m^3 cube in the soil (exact dimensions were 0.25m^2 surface area, dug 0.5m deep). The soil in the plot was removed and shaken through the screen sieve. Rocks were removed and earthworms were collected. Information about the earthworms was recorded.

Student 1

Student 1 collected the information about the length of each earthworm. This information is shown in Table 1.

Table 1		
Plot #	Average Length of Earthworm (cm)	
	Field	Forest
1	2.20 cm	3.10 cm
2	10.3 cm	0.91 cm
3	4.05 cm	1.60 cm
4	8.90 cm	1.90 cm
5	7.60 cm	2.04 cm

Student 2

Student 2 collected the information about the mass of each earthworm. This information is shown in Table 2.

Table 2		
Plot #	Average Mass of Earthworms (g)	
	Field	Forest
1	1.10 g	12.3 g
2	4.30 g	9.50 g
3	5.15 g	14.4 g
4	1.90 g	13.4 g
5	1.75 g	16.0 g

16. Which graph best describes the size of earthworms found in a forest?

F

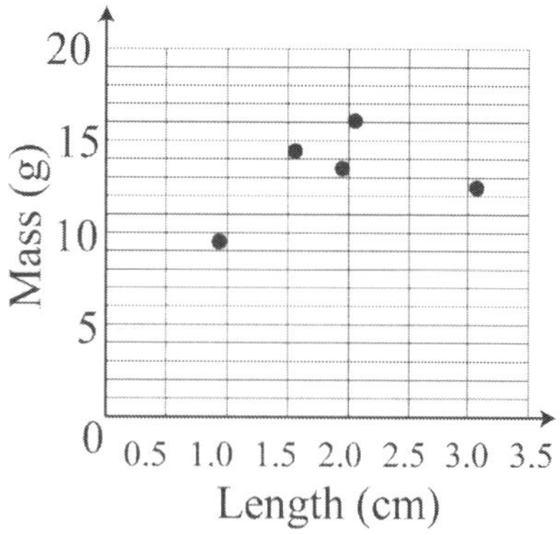

G

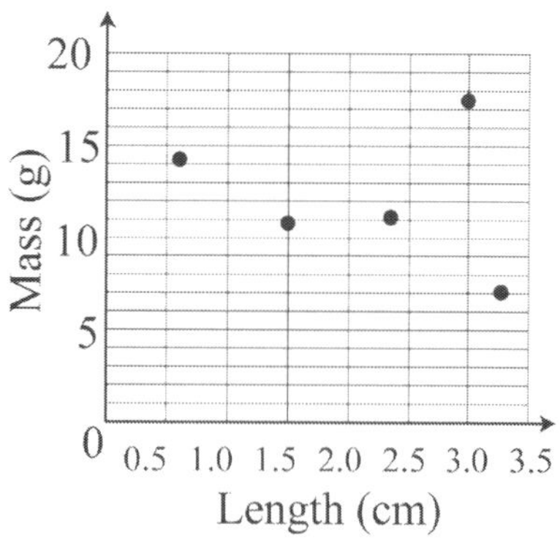

H

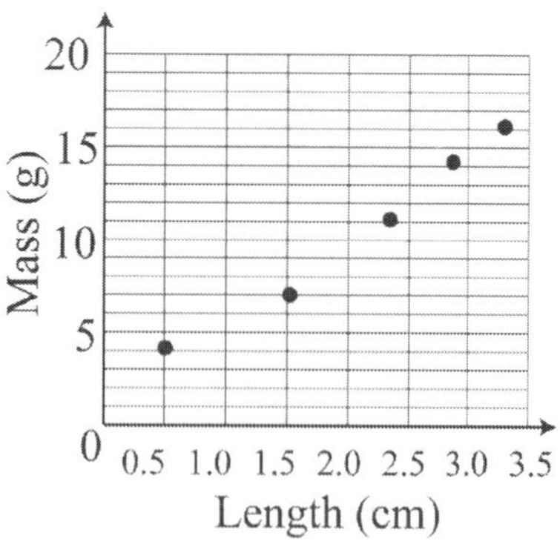

J

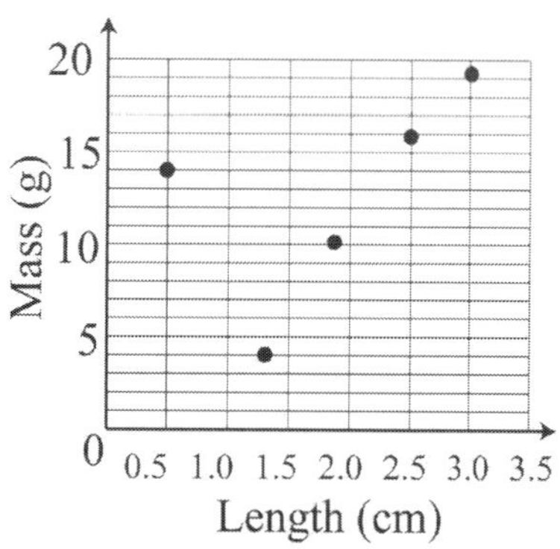

17. According to Table 1, which field plot contained the longest earthworms?

A 1

B 2

C 3

D 4

18. Which statement below best summarizes the trends seen in this study?

F Although the earthworms in the field were longer, the earthworms found in the forest were more massive.

G Earthworms found in the field were shorter but more massive.

H The average length of earthworms in the forest was most similar between plots 1 and 2.

J There is no difference in size or length among earthworms found in the forest versus earthworms found in the field.

19. While working in a forest plot, which earthworm measurement below would be most reasonable?

A 3.01 cm

B 7.90 cm

C 9.39 cm

D 27.4 cm

20. During the plot selection process, why were plots with similar slope and amounts of sunlight selected?

F To ensure predatory activities remain consistent between the two areas

G To ensure the differences in earthworms were due to differences in soils

H To prevent sunlight from drying the earthworms

J To prevent differences in temperature from affecting earthworm activity

21. What was the main use of the sieve?

A To remove rocks

B To remove water

C To isolate the earthworms

D To weigh the earthworms

Passage V

Wildfires are beneficial for forest regeneration. Land management may influence the severity of wildfires. *Salvage logging* is the removal of dead or dying forest trees following a fire or disease outbreak. Some say salvage logging decreases the risk and severity of future fires, while others argue it increases the risk and severity of fires. A forest manager and an ecology professor discuss whether salvage logging is a viable land management practice.

Forest Manager

Dead or diseased trees serve as the major fuel for future fires. Therefore, the best way to minimize the risk of fire is to remove the dead timber from the forest floor. Salvage logging followed by replanting ensures the reestablishment of desirable tree species. For instance, planting conifers accelerates the return of fire-resistant forests.

The black surfaces and ash layers of burned landscapes result in very high soil temperatures that are lethal for many plant species. Logging mixes the soil, thereby decreasing surface temperatures. After an area has been salvage logged, the area shows regeneration occurring almost immediately. Regeneration can take several years in unmanaged areas.

Ecology Professor

Salvage logging removes large, less-flammable woody material and leaves behind small dead limbs and branches. This process transfers small, broken branches and other material to the forest floor where it is available as fuel and increases the risk of fire severity.

Reforestation is essential after a fire or infestation. However, replanting of a single type of tree creates monoculture plantations. When there is low species diversity, the chance that disease and insect infestation will weaken the forest is increased. Salvage logging also disrupts the soil, causing more erosion and sedimentation. It removes shade needed for young hardwood seedlings to grow.

22. A study compared two plots of land that were managed differently between two fires that occurred 15 years apart. After the first fire, plot A was salvage logged and replanted with conifers, and plot B was left unmanaged. They performed a controlled burning of both plots and found that plants burned with greater severity in plot A than in plot B. Which viewpoint do these results support?

F Only the manager

G Only the professor

H Both the manager and professor

J Neither the manager nor the professor

23. Woody materials left behind from a wildfire can serve as fuel for future fires. According to the forest manager, how should these materials be dealt with?

A They should be left on the forest floor after a wildfire, providing nutrients for the soil.

B They are left on the forest floor after a wildfire to serve as shade for hardwood seedlings.

C They should be removed from the forest floor.

D They should be distributed more evenly across the forest floor after a wildfire.

24. A new study analyzing the negative effect of bore beetle infestation in a monoculture oak plantation could help support and validate the arguments of which viewpoint?

F Only the manager

G Only the professor

H Either the manager or professor

J Neither the manager nor the professor

25. According to the professor, younger forests are more susceptible to increased fire severity than old growth forests. Which of the following statements does NOT support this point?

A In younger forests, branches are closer to the forest floor and more available for fires.

B Old growth forests contain larger and taller trees, where branches are high and unavailable to fires.

C Younger forests have less biological diversity and less disease-resistant species.

D Young forests contain more disease-resistant species, and large trees common in old growth forests serve as the main fuel source for fires.

26. If salvage logging continues, which of the following would most likely result according to the manager?

- **F** Young, conifer forests and an abundance of fire-resistant tree species
- **G** Young, diseased tree species with eroded soil
- **H** Old growth forests with small, woody materials on the floor
- **J** Young forests with woody material on the floor

27. Which of the following statements would be supported by both the manager and professor?

- **A** Large, less flammable timber should be left unlogged after an infestation.
- **B** Reforestation is an essential component of forest regrowth.
- **C** Replanting conifer trees decreases the vulnerability of forests to future fires.
- **D** Salvage logging is beneficial only after a very severe wildfire.

28. What facts does the manager use to support his claim that salvage logging is beneficial?

- **F** Decreased soil surface temperatures and immediate forest regeneration
- **G** Lower forest biodiversity and increased risk of plant disease and disruption of soils
- **H** Increased soil erosion and sedimentation
- **J** Wildfires are influenced by water availability, not land management practices

Passage VI

Atoms of the same element with different numbers of neutrons are called isotopes. Many isotopes are unstable and undergo a process called *nuclear decay.* This is a process by which unstable isotopes emit radiation, becoming a more stable element. Isotopes that undergo nuclear decay are said to be radioactive and are sometimes called radioisotopes.

Different isotopes decay at different rates. The amount of time necessary for one half of the atoms in a radioisotope sample to decay is called that radioisotope's half-life. Two researchers were asked to quantify the decay of the three radioisotopes in Table 1.

Researcher 1

As radioisotopes decay, the level of radiation emitted by a sample drops proportionally. Researcher 1 used this fact to approach the project mathematically. He calculated the percentage of undecayed radioisotopes remaining after a given number of half-lives. This is shown in Figure 1.

The result was a single mathematical model that describes the behavior of any radioisotope.

Researcher 2

Researcher 2 decided to compile raw data on the masses of the isotopes left after certain intervals. He began with a 50 milligram sample of each of the three radioisotopes in Table 1. His data is graphed in Figure 2.

Table 1

Isotope	Symbol	Half-life
technetium-96	Tc-96	4.3 days
berkelium-246	Bk-246	1.8 days
radon-222	Rn-222	3.8 days

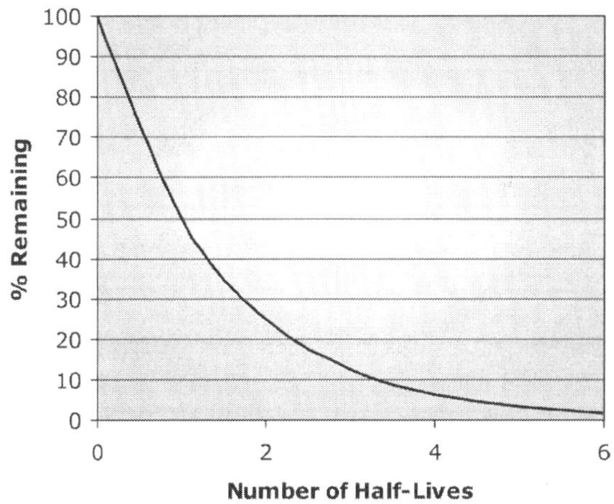

Figure 1

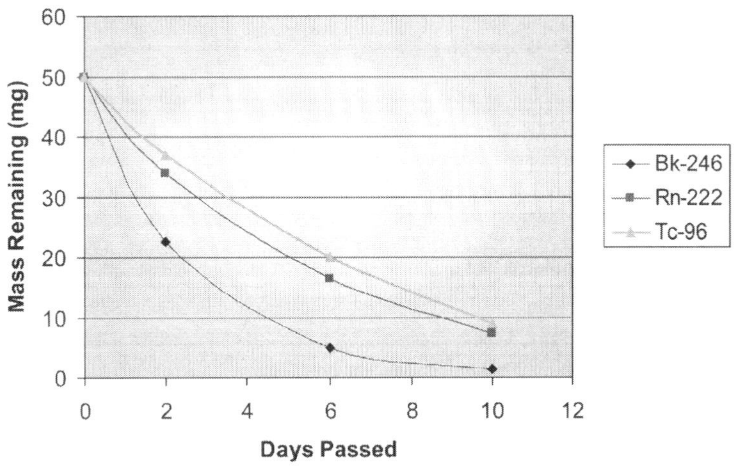

Figure 2

29. Using Figure 1, if you started out with a sample of 1,000 radioactive atoms, how many atoms will have decayed after 2 half-lives?

A Around 750

B Around 500

C Around 250

D Around 125

30. After 9 days, Bk-246 has gone through 5 half-lives. Using Figure 1, what percentage of the original 200 gram sample can be expected?

F 25%

G 10%

H 3%

J 0%

31. According to Figure 2, 20 mg of a radioisotope were present after 6 days. Which radioisotope is that?

A Technetium-96

B Berkelium-246

C Radon-222

D None of them

32. The researcher next obtained a sample of one radioisotope from Table 1. He isolated the sample in a lead box. When he returned in a month, he found his sample almost entirely changed into another element. Which of the isotopes in Table 1 did the researcher most likely have?

F Technetium-96

G Berkelium-246

H Radon-222

J All of these samples would have almost entirely decayed in a month.

33. If you did not know the half-life of the three radioisotopes in Figure 2, how could you tell which one had the longest half-life?

A It is the one that has the most mass remaining.

B It is the one that has the least mass remaining.

C It is the one that started out with 50 mg.

D It is the one with the fastest decay rate.

34. Researchers 1 and 2 took different paths to reach the same conclusions. Another scientist wants help characterizing the decay patterns of a new radioisotope, radon-211. This radioisotope has a half-life of 14.6 hours. Which researcher's data can be quickly used to predict the amount of radon-211 left after 3 half-lives?

F Researcher 1 only

G Researcher 2 only

H Researchers 1 and 2

J Neither researcher

Passage VII

In the summertime, large cities are often plagued with *smog*. Smog is brown-colored haze that can irritate the nose, lung, eye and throat tissue. It is composed of many chemicals, including nitrogen monoxide gas (NO), nitrogen dioxide gas (NO_2) and ozone (O_3). Ozone is known to damage human tissues. Table 1 shows the effects various concentrations of ozone have on humans. In high concentrations, smog can cause serious health problems and even death, especially in the elderly, small children and individuals with heart and lung disease.

The development of smog is a multi-step process. Nitrogen monoxide is a pollutant in car exhaust. The chemical formula for nitrogen monoxide is (NO). First, the nitrogen monoxide released by cars reacts with the oxygen (O_2) in the air to produce the reddish-brown gas nitrogen dioxide (NO_2). Then, sunlight (hv) starts a process by which nitrogen dioxide gas reacts with oxygen gas (O and O_2) to produce ozone. These processes are shown in chemical Reactions 1 and 2.

$$2NO + O_2 \longrightarrow 2NO_2 \quad \text{Reaction 1}$$

$$NO_2 + O + hv \longrightarrow NO + O_3 \quad \text{Reaction 2}$$

Researchers studying the levels of smog in a city took several measurements. First, they recorded the average levels of nitrogen monoxide and ozone as they fluctuate during and after the morning commute. The data from this study is displayed in Figure 1. They then recorded the daily noontime ozone concentrations over the course of a two-week period in July. The results from this study are displayed in Figure 2. Table 2 provides weather data recorded over the same two-week period, from July 1 through July 14.

Table 1

Ozone concentration (ppm)	Effects and Recommendations
<0.164	No noticeable effects
0.164 – 1.64	All persons may notice some degree of lung irritation. Those at high risk for lung or heart problems should stay indoors and reduce physical activity.
>1.64	Eyes, nose and throat irritation will be widely noticed. Breathing difficulties are likely. All should avoid outdoor activity.

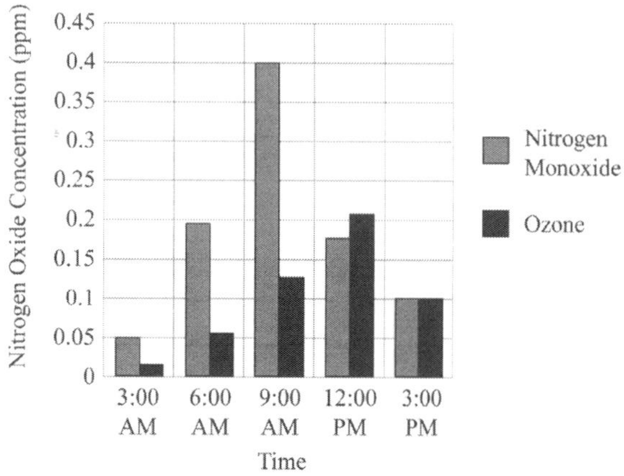

Figure 1

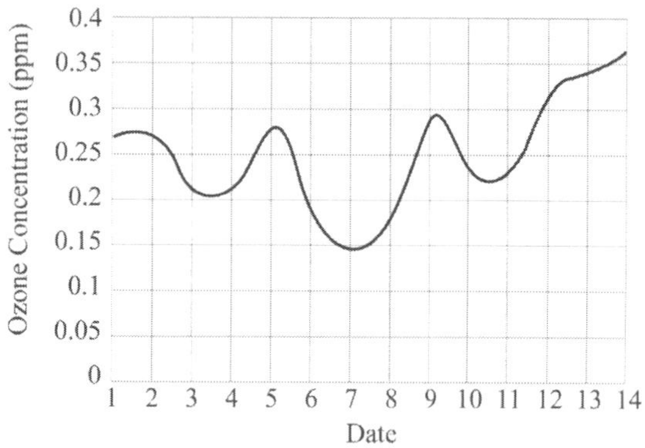

Figure 2

Table 2		
Date	High Temperature (degrees Fahrenheit)	Conditions
1-Jul	85	Sunny
2-Jul	89	Sunny
3-Jul	87	Partly Cloudy
4-Jul	86	Sunny
5-Jul	83	Overcast
6-Jul	81	Rainy
7-Jul	79	Rainy
8-Jul	77	Sunny
9-Jul	75	Sunny
10-Jul	84	Sunny
11-Jul	87	Sunny
12-Jul	86	Partly Cloudy
13-Jul	89	Partly Cloudy
14-Jul	91	Mostly Cloudy

35. Use Reactions 1 and 2 to determine the chemical formula for ozone.

A NO

B O

C O_2

D O_3

36. Using Figure 1, which of the following best describes the relationship between nitrogen monoxide concentration and ozone concentration?

F Nitrogen monoxide reaches a maximum about 3 hours after ozone.

G Increasing ozone results in increasing nitrogen monoxide.

H Increasing nitrogen monoxide is followed by an increase in ozone.

J No relationship seems to exist.

37. According to Table 1 and Figure 1, at what times should high-risk individuals remain indoors?

A Around 7 a.m.

B Around 9 a.m.

C Around noon

D Late in the afternoon

38. Using Figure 2 and Table 2, what effect does the weather appear to have on the formation of smog?

F Rain seems to accelerate the formation of smog.

G Rain appears to reduce the incidence of smog.

H Sunny weather appears to inhibit the formation of smog.

J Clouds seem to prevent smog from forming.

39. Using Table 1 and Figure 2, during which dates of the study did ozone concentrations remain in a range that resulted in no noticeable health effects?

A July 5 and 9

B July 3, 4, 6 and 7

C July 6 and 7

D July 3 and 10

40. On weekends, fewer cars are driven. Use Figure 2 and Table 2 to hypothesize which days during this study fell on the weekends.

F July 4 and 5, July 9 and 10

G July 3 and 4, July 6 and 7

H July 3 and 4, July 10 and 11

J July 6 and 7, July 13 and 14

ACT Science Reasoning Test Evaluation Chart

Directions: Use the chart below to better focus your study for the ACT Science Reasoning Test. Identify the question range or questions you answered incorrectly. Turn to the appropriate chapters, read the explanations and complete the extra exercises. Review other chapters as needed. Review other chapters as needed. Finally, complete the Practice Test 2 as final preparation for the ACT Science Reasoning Test.

Chapter #	Question #
Chapter 2: Data Representation	1–5; 6–10; 11–15
Chapter 3: Research Summary	16–21; 29–34; 35–40
Chapter 4: Conflicting Viewpoints	22–28

z

Index